Donald D Harrington Symposium on the Geology of the Aegean 2008

IOP Conference Series: Earth and Environmental Science Volume 2

Austin, Texas, USA
28 – 30 April 2008

ISBN: 978-1-5108-0148-6
ISSN: 1755-1307

Printed from e-media with permission by:

Curran Associates, Inc.
57 Morehouse Lane
Red Hook, NY 12571

Some format issues inherent in the e-media version may also appear in this print version.

Printed by Curran Associates, Inc. (2015)

For permission requests, please contact the Institute of Physics
at the address below.

Institute of Physics
Dirac House, Temple Back
Bristol BS1 6BE UK

Phone: 44 1 17 929 7481
Fax: 44 1 17 920 0979

techtracking@iop.org

Additional copies of this publication are available from:

Curran Associates, Inc.
57 Morehouse Lane
Red Hook, NY 12571 USA
Phone: 845-758-0400
Fax: 845-758-2634
Email: curran@proceedings.com
Web: www.proceedings.com

TABLE OF CONTENTS

The Aegean: A natural laboratory for tectonics

B Clark Burchfiel[1]
Department of Earth, Atmospheric, and Planetary Sciences, Massachusetts Institute of Technology, Cambridge, MA 02139, USA

E-mail: bcburch@mit.edu

Abstract. The Aegean, a young and active tectonic region, is a natural laboratory for analyzing many tectonic processes that occur in backarc extensional regimes, and the correlation of these processes from landscape development to deeper mantle dynamics. Cenozoic development of the Aegean region was dominated by subduction beneath Europe and coeval upper plate extension modified by westward extrusion of Anatolia. Intraorogenic and backarc extension began during early Cenozoic time within the Balkans and NW Turkey during closure of the Vardar ocean. Extension was manifested by core complex formation and a change in volcanism caused by the evolution of the lithosphere and mantle wedge. Following a short period of local (?) shortening in ~ early Miocene time, regional extension began and continued to the present. Within the Hellenides, E-W extension and the subduction zone migrated westward as thick and thin crustal units were progressively accreted and were complexly rotated up to 40° CW. Within the eastern Balkans and NW Turkey, N-S extension migrated westward and southward, and in the Aegean the volcanic arc and subduction zone migrated southward. Turkish crustal elements rotated complexly CCW, which in concert with the CW rotation in the Hellenides increased the curvature of the subduction zone and lengthened the orogen causing greater subsidence and extension in the Aegean Sea. Westward extrusion of Anatolia from the Arabian collision zone was enhanced by slab roll back in west moving Aegean crust more rapidly westward. Abundant evidence supports slab rollback at different velocities along the subduction zone. In Pliocene time, the North Anatolian fault crossed the Hellenides in a complex transtensional zone and a diffuse zone of left-lateral shear crossed western Turkey at present isolating a relatively undeforming Aegean plate. Major tectonic questions include: What is the geometry and fate of subducted slabs?, How much crust is accreted during subduction of thick and thin crust?, Does crustal accretion cause slab break off or only changes in slab dip?, How does the mantle wedge evolve and how is this expressed in the volcanism?, How much lateral mantle flow occurs during changes in slab geometry and rotation of crustal elements?, and What is the relation of mantle and crustal structure to first order landscape development?

1. Introduction
Active orogens, such as the Aegean region (figure 1), are natural laboratories for analyzing many tectonic processes that occur in backarc extensional regimes, and the correlation of these processes from landscape development to deeper mantle dynamics. The Cenozoic development of the Aegean region was dominated by subduction beneath Europe and coeval upper plate extension modified by westward extrusion of Anatolia beginning in Late Miocene time. Mesozoic events took place within

[1] To whom any correspondence should be addressed.

the Vardar oceanic track between Europe and the Pelagonian, Sakar, and Anatolian-Tauride continental fragments, and possibly within the Pindus ocean.

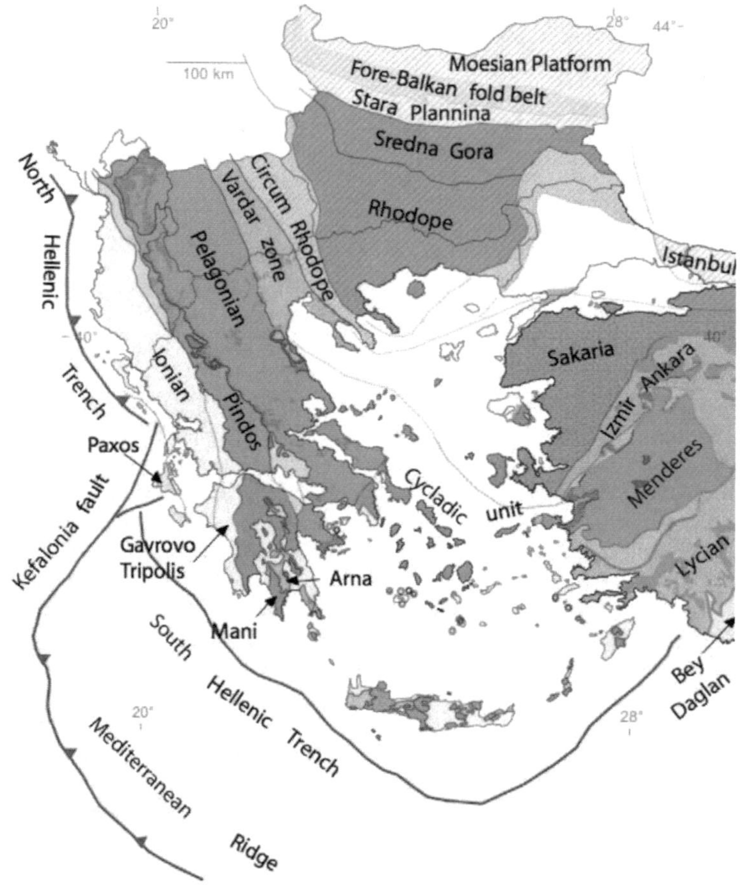

Figure 1. Major structural elements of the Aegean Sea region.

The events considered here begin with the closure of the Vardar, IntraPontide and Izmir-Ankara oceans in about Middle Eocene time and are mainly manifested in the southernmost Balkan Peninsula and NW Turkey. During the final closure of the oceanic tracks, there were many important features developed that are related to the dynamics of systems that are transitional from end of northward subduction to a subduction that migrated into the Pindus deepwater paleogeographic unit and south of the Menderes massif in Turkey. During the early part of the transition, both ductile thrusting and extensional low-angle structures have been described in deep exposures of curst commonly referred to as core complexes. Coeval with these structures an E-W fold thrust belt was active in north-central Bulgaria suggesting the region from the oceanic tracks in the south to central Bulgaria was in a regional compressional environment. The ductile low-angle structures are probably best interpreted as ductile flow in a thick hot crust. By Late (possibly early Middle) Eocene time SW-NE to E-W extension on low-angle faults began in western Bulgaria and continued into Oligocene time. These west-dipping low-angle faults may have begun E-W extension in the western Balkans that continued to the present migrating from western Bulgaria to central Albania, although there may have been a short period of local compression in late Oligocene to early Early Miocene time. In central and eastern Bulgaria, and possibly extending into NW Turkey, extension was associated by abundant volcanism that shows a progressive of thinning of lithosphere in the mantle wedge. N-S extension produced narrow half grabens that were developed not only in the area of volcanism, but also in a nonvolcanic setting farther north just south of the inactive fold-thrust belt in the Forebalkan of north central Bulgaria. The changes structure and volcanism suggest the subducted slab, which began as oceanic lithosphere, changed to continental lithosphere and subduction slowed and slab dip steepened thinning the lithosphere and causing extension in the overriding plate.

Closing of the Vardar and Izmir-Ankara oceans produced different results in Turkey and in Greece; in Turkey a large track of continental crust (the Anatolide-Tauride track) was accreted whereas in Greece a less wide piece (the Pelegonian) was added. In Turkey an extensive sheet of oceanic and continental margin rocks (Lycian nappes) was emplaced above the Anatolide-Tauride fragment. Shortly after their emplacement N-S extension began above this fragment exposing the metamorphic rocks of the Menderes massif by Oligocene time. This exposure of high-grade metamorphic rocks is often referred to as a core complex, but it is more likely that it was elevated by continued underthrusting of continental lithosphere beneath it as the Lycian nappes were moved farther south. In Greece and Albania coeval Late Eocene –Oligocene thrusting subducted the crust (thin continental or locally oceanic?) beneath the Pindus deep-water rocks forming closely spaced series of folds and thrust. This thrusting began a rather continuous subduction of crust below tectonic units characterized by thick (Gavrovo-Tripolis and Paxos units) and thin (Pindos, Ionian and present Ionians sea floor) crust. Beginning in Late Miocene, coeval with thrusting, the thrust sheets developed on the Greek-and Abanian mainland rotated more than 40° CW with a pivot point in northern Albania. This rotation and the extension that followed behind the rotating thrust sheets is commonly ascribed to trench roll back, however, two of the important tectonic problems for the Hellenides is, if trench roll back occurred, what makes up the crust below the thin stacked nappe sheets in the Hellenides and how much of the crust or lithosphere is involved in the Late Miocene to Present rotation?

During nappe emplacement, there are two belts of HP-LT units emplaced into the thrust pile; a northern one that was emplaced in Eocene time and has yielded metamorphic ages that range from 70 to 40 Ma, and a southern one that yields late Oligocene-Early Miocene ages. These units usually show unroofing by normal faults at the top and thrusting at their base. It has been suggested that they are emplaced by oblique extrusion of these rocks into the nappe pile, so that rock below and above were never subjected to metamorphic conditions comparable to the HP-LT units. The older belt units in Turkey lie above the Menderes metamorphic rocks but in southern Greece they lie below the Pelegonian unit. Not uncommonly these two units are correlated, but the presence of the HP-LT units I these two different structural positions suggests otherwise and needs to more carefully examined.

The migration of thrusting in the Hellenides and Taurides and the rotation in the Hellenides has caused the two thrust systems to diverge, thus stretching and thinning the crust in the region between them, the Aegean Sea. Volcanism also shows southward asymmetric migration across the Aegean. Late Oligocene-Miocene volcanism migrated south from the Balkans and occurred was extensive in NW Turkey and central Aegean, but is scattered and rare in the Hellenides. Pliocene-Quaternary volcanism occurs in a narrow belt in the central Aegean extending into western Turkey, but is rare along the Hellenides.

The simple system of trench rollback changed in Late Miocene time by the westward extrusion of Anatolia from the Arabian/Turkish collision zone in the east. The extrusion was further facilitated by trench rollback as GPS data shows the Aegean at present is moving faster to the SSW than Anatolia is moving west, thus extension occurs above the subduction zone. At present the Aegean is moving more or less as an undeformed plate bounded on the north by the North Anatolian fault zone and its broad western extension though the Corinth region, and on east by a broad diffuse zone of left-lateral shear in western Turkey characterized by a series of ~E-W tending grabens. The geology of the region suggests this pattern developed in Pliocene time when the North Anatolian fault zone cut across southern Greece.

The formation of the North Anatolian fault zone mostly disconnected extension in the southern Balkans from extension in the Aegean. In Bulgaria N-S extension begins in the Late Miocene gradually extensions west, as the North Anatolian fault extends west, into a region in which E-W

IOP Conf. Series: Earth and Environmental Science **2** (2008) 012001 doi:10.1088/1755-1307/2/1/012001

extension coevally migrates across western Bulgaria into central Albania; the N-S extension in Bulgaria can be interpreted to be the result of slow movement of the lithosphere drawn southward by the SSE rapid motion of the Aegean plate. The westward migration of E-W extension in the SW Balkans can be interpreted as continued trench roll back with slowing and steepening of the subducted slab because of attempted subduction of the continental crust of the below the Adriatic.

2. Key questions

The Aegean region contains an active system ripe for studying many exciting tectonic and dynamic problems of both local and global importance. There are differences of many features from western Turkey to mainland Greece still remain to be elucidated, and include the correlation of tectonic elements between Turkey and Greece, the difference in the abundance and evolution of volcanism across the Aegean and the difference in tectonic evolution of the Hellenides and Taurides. More globally applicable studies are numerous. They include:

1. The need to understand how the middle and lower crust is generated in a slab roll back regime,
2. The dynamics of the emplacement and modification of HP-LT structural units,
3. The processes of lower crustal / lithospheric deformation and modification through time - high resolution geophysics,
4. The evolution of volcanism and processes in mantle wedge dynamics in shortening, extensional and transitional settings,
5. What events do geophysical data, such as shear wave splitting and tomographic images record,
6. Depth of rotation recorded by paleomagnetic data,
7. The extent of decoupling between upper crustal structure and deeper crust and mantle dynamics, particularly in this region where the great thinness of the thrust sheets can be clearly demonstrated,
8. What is the fate of subducted slabs and does accretion of continental crustal material cause slab break off or only changes in slab dip,
9. How much lateral mantle flow occurs during changes in slab geometry and rotation of crustal elements, and
10. What is the relation of mantle and crustal structure to first order landscape development?

Archaeological elements of Mt. Lykaion Sanctuary of Zeus (southern Peloponnesus) in relation to tectonics and structural geology

George H Davis[1]
Regents Professor, The University of Arizona, Department of Geosciences, 326 Gould-Simpson Building, Tucson, AZ 85721 , USA

E-mail: gdavis@u.arizona.edu

Abstract. The Sanctuary of Zeus is the focus of the Mt. Lykaion Excavation/Survey (University of Pennsylvania, University of Arizona, and 39th Ephorate of Prehistoric and Classical Antiquities). It was described by Pausanias as a sacred place of pan-Hellenic significance, with stadium and hippodrome in which athletic games were held, a sanctuary of Pan, and a formidable temenos and altar of Lykaion Zeus. In picturing human activity on this mountain during ancient times, it is not adequate to treat the mountain as if it were simply a tall, symmetrical, and handy edifice within which rock contents are irrelevant, for the geology *within* Mt. Lykaion significantly influenced what was built *on* it, and where! There are contemporary reminders of the 'power' of the site, including the devastating April, 1965, Megalopolis earthquake, the epicenter of which was merely 4 km away. In fact, there are active normal faults within the sanctuary. However the primary geoarchitecture is that of the Pindos fold and thrust belt, fashioned largely in Cretaceous through Eocene. Mt. Lykaion's dome-like summit is a thrust klippe separated from underlying nappes by a major thrust fault (Lykaion thrust), the subhorizontal trace of which encircles the mountain creating a subtle bench in the landscape coinciding closely with archaeological and natural elements important to the sanctuary (e.g., stoa, seatwall, fountains, trails). Late Jurassic through Eocene "Pindos Group" formations are stacked and repeated by the thrusting. Inter-relationships between bedrock, structure, and archaeology are revealed in a "geoarchaeological column," which displays positioning of elements in relation to the thrust, and orientations of rock formations in relation to flat patches in otherwise steep, rocky country, which became sites suitable for placement of hippodrome, baths, temenos, horse pasturing areas, etc. Worked limestone blocks are locally derived and can be matched with formations. The compelling high elevation of the ash altar is testimony to residual crustal buoyancy achieved through "Pindos" crustal shortening, and the steep processional ascent to the altar speaks to regional active normal faulting and rapid erosion.

1. Introduction

Mt. Lykaion, Sanctuary of Zeus (elevation ~1400m), resides in Arcadia, approximately 30 km southeast of Olympia and ~10 km west of Megalopolis (figure 1).

[1] To whom any correspondence should be addressed.

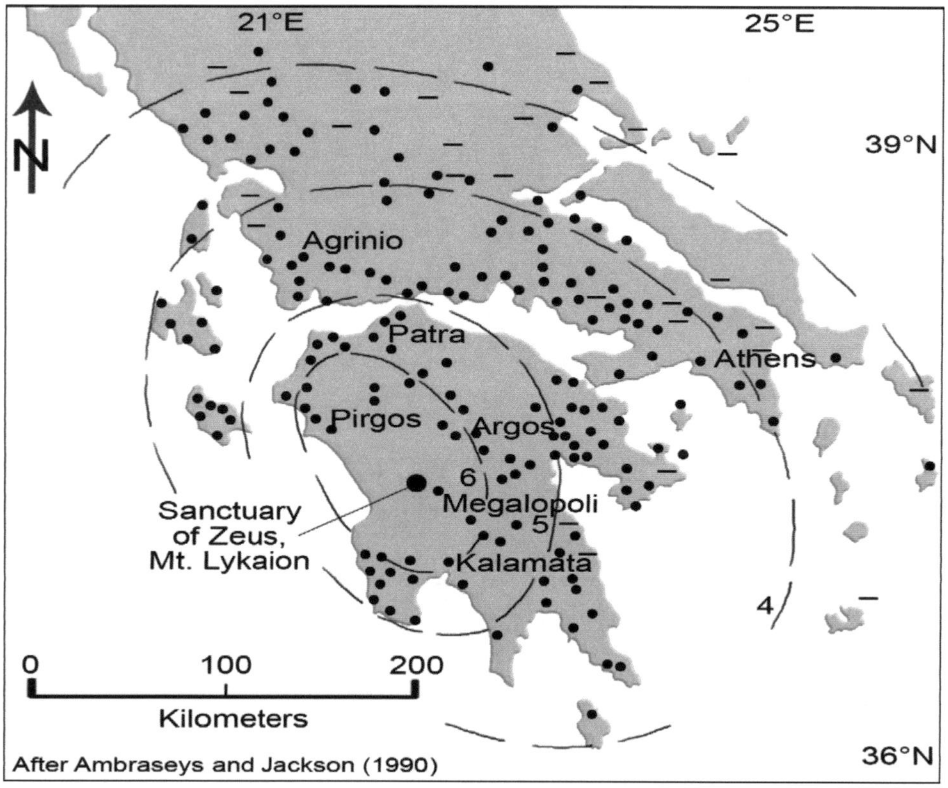

Figure 1. Location of Sanctuary of Zeus, Mt. Lykaion. The base map [from 1] shows the area of influence of the April, 1965 Megalopolis earthquake, the epicenter for which was only 4 km from the sanctuary. Solid black dots refer to locations for which there were sufficient information to estimate intensity.

This locale is the object of an interdisciplinary Greek-U.S. partnership engaged in surveying and excavating of the archaeology of the Sanctuary of Zeus, Mt. Lykaion, in the Peloponnesus http://lykaionexcavation.org/. Participating institutions are The University of Pennsylvania, The University of Arizona, the 39th Ephorate of Prehistoric and Classical Antiquities (Sparta), the American School of Classical Studies (ASCSA), and the Archaeological Institute of Peloponnesian Studies. Co-Directors Dr. David Romano (University of Pennsylvania Museum) and Dr. Mary Voyatzis (Department Head of Classics, University of Arizona) invited me into the project. My role is one of examining potential geological influences on the positioning and presence of built structures within the sanctuary, and the geological control of certain natural, utilized phenomena (e.g., springs). In the 2nd century B.C., the Sanctuary of Zeus was described by Pausanias [see 2] as a sacred place of pan-Hellenic significance, with stadium and hippodrome in which athletic games were held, a sanctuary of Pan, and a temenos and altar of Lykaion Zeus (Pausanias, 8.38.5; 8.38.6; 8.38.7). The following elements are found in the lower part of the sanctuary: the hippodrome and stadium, baths, xenon, seating complex, stoa, processional way (?), springs (including the Agno Fountain), and horse pasturing area (?). The elements in the upper part of the sanctuary include the early stadium (?), column bases, temenos, and ash altar [3] (figure 2).

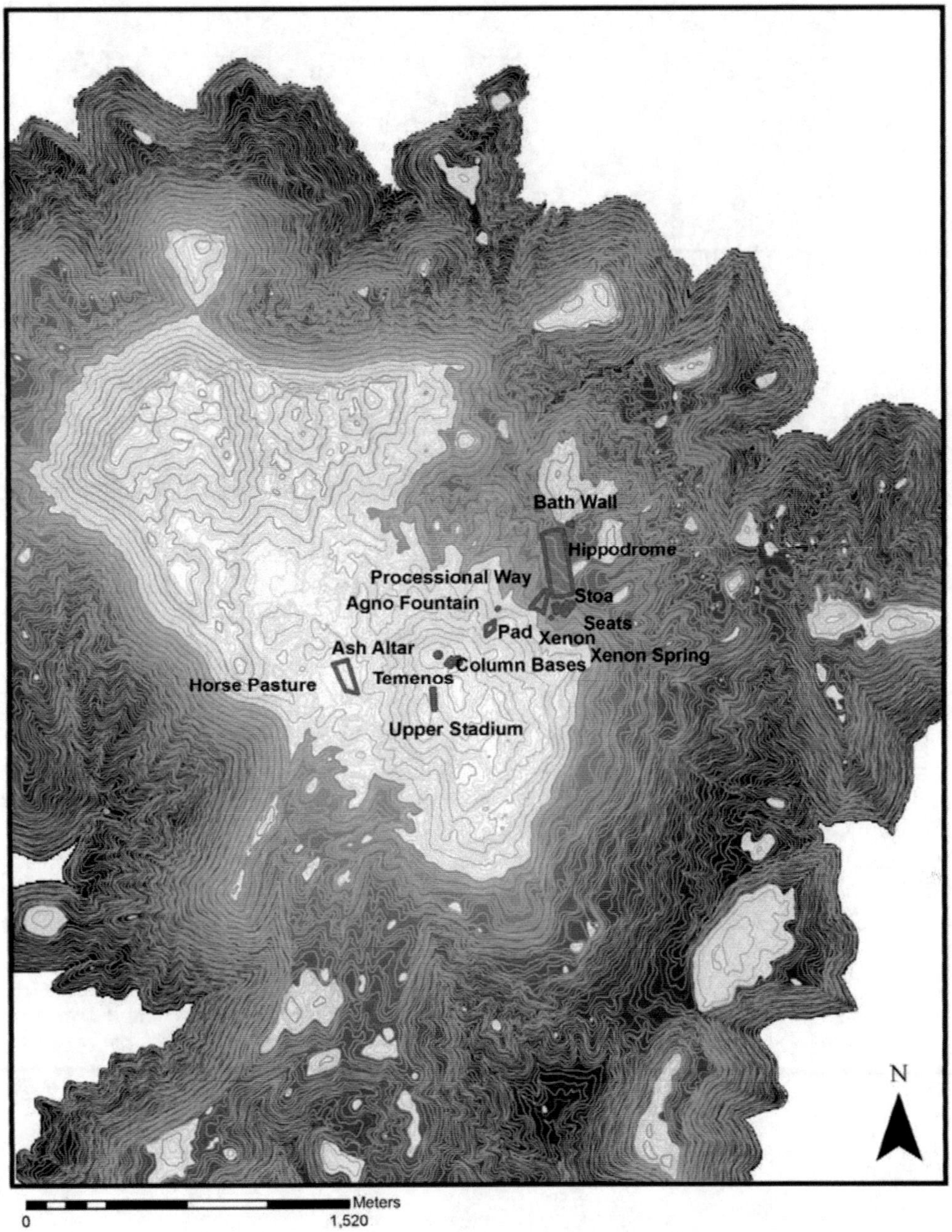

Figure 2. Map distribution of archaeological elements, Sanctuary of Zeus, Mt. Lykaion. Contour interval of topographic base map is 4 m. White areas are highest elevations. Dark areas give a sense for how the summit of the mountain "drops off."

In picturing human activity on this mountain during ancient times, it is not adequate to treat the mountain as if it were simply a tall, symmetrical, and handy edifice within which rock contents are irrelevant (figure 3), for the geology *within* Mt. Lykaion significantly influenced what was build *on* it, and *where*!

Figure 3. Mt. Lykaion, as viewed from the northwest. Summit is surmounted by the ash altar.

Though the Sanctuary of Zeus has received the attention of scholars, the site has not been investigated for ~100 years. At the turn of the 19th century K. Kourouniotes [e.g., 4] carried out reconnaissance trenching and surveying, and he identified the area of the temenos and the features of the lower sanctuary. He uncovered bronze figurines (mainly portraying Zeus) close to the column bases near the edge of the temenos. He opened several trenches at the ash altar, determining that the ash layer was ~1.5m thick, and documenting the presence of ash, rocks, bone, pottery, roof tiles, coins, and bronzes, including early 7th century BC bronze tripods (figure 4).

There is authoritative ancient literary and historical evidence that informs us about Mt. Lykaion (see http://lykaionexcavation.org/), especially the descriptions provided by Pausanias based upon his visit to the site in the 2nd century AD. For example, according to Pausanias (8.38.7, 8.38.6) [2]:

> "*On the highest point of the mountain is a mound of earth, forming an altar of Zeus Lykaios, and from it most of the Peloponnesus can be seen. ... On this altar they sacrifice in secret to Lykaion Zeus.*" ..."*Among the marvels of Mount Lykaion the most wonderful is this. On it is a temenos of Lykaion Zeus, into which people* [except for priests] *are not allowed to enter. If anyone takes no notice of the rule and enters, he must inevitably live no longer than a year.*"

Figure 4. Trench excavation of ash altar, Sanctuary of Zeus, during the summer of 2007. Archaeologist is Dr. Arthur Rohn.

2. Stratigraphy

Sedimentary rocks of the Pindos Group form the bedrock of Mt. Lykaion Sanctuary of Zeus. Pindos Group stratigraphy is a product of late Triassic through Eocene deposition in a deep, narrow, sediment-starved basin known as the Pindos Basin. Lithologies comprising the Pindos Group include radiolarian cherts, turbidites, and flysch deposits (Jurassic), succeeded upwards by a thick limestone succession (Upper Cretaceous), and topped with flysch (Paleocene-Eocene) [5, 6, 7, 8, 9, 10, 11]. Rocks of the Pindos Group comprise a major lithotectonic assemblage forming a 500km-long belt, averaging ~50km in breadth (i.e., the Pindos Zone) that trends N-S through the whole of Greece (figure 5). The formations, which I mapped (figure 6) are indeed part of the Pindos Group. The Pindos Zone is an intensively deformed fold-and-thrust belt marked by E-W compressional shortening exceeding 50% [11] (figure 7). The folding and thrusting is a product of inversion tectonics, for in the early Tertiary the Pindos Basin began to be closed by compressional tectonics. The Triassic through Upper Cretaceous Pindos Group sediments were progressively uplifted, becoming weathered and eroded when they rose above sea level. Such cannibalization led to deposition of the Paleocene-Eocene flysch into the closing and shallowing Pindos Basin. In turn these newly minted sediments became folded and thrusted.

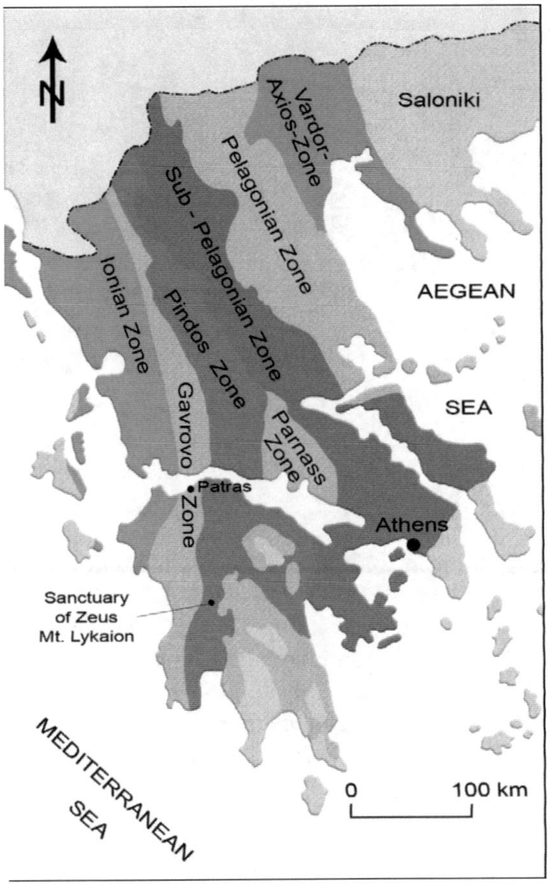

Figure 5. Map of regional distribution of the Pindos Zone lithotectonic assemblage (i.e., Pindos Group stratigraphy) (after [12]).

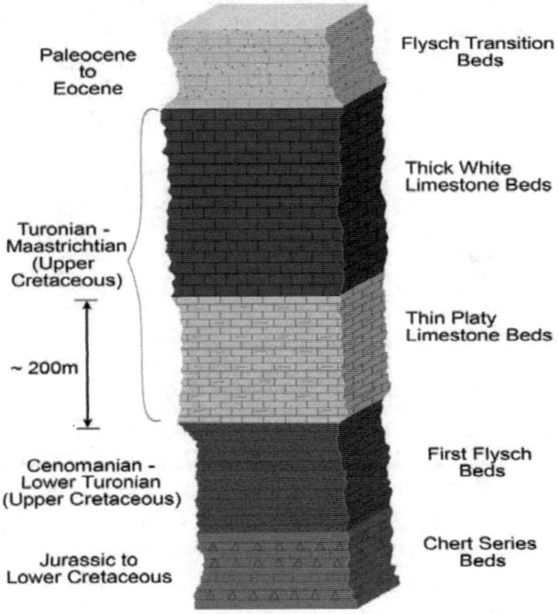

Figure 6. Schematic portrayal of the stratigraphy of the Sanctuary of Zeus, Mt. Lykaion. These are the fundamental units defined for the purpose of the 1:5000-scale geologic mapping carried out by George Davis.

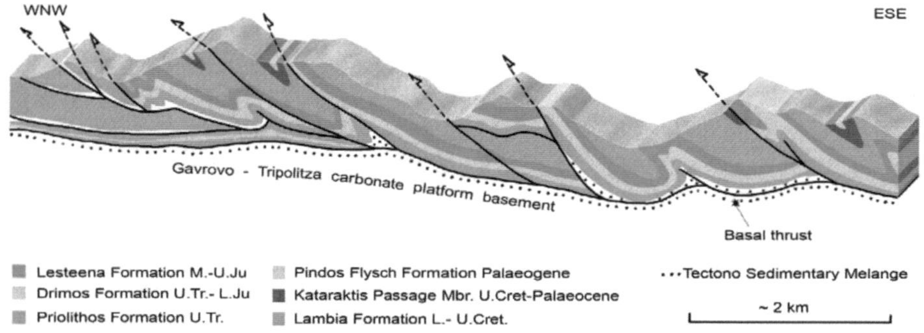

Figure 7. Nature of folding and thrusting within the Pindos Zone. Compressional shortening generally exceeds 50%. After [8].

Figure 8. Outcrop appearance of Chert Series Beds. Ribbon cherts and thin limestone beds have distinctive expression. Folding and shearing is abundant within the Chert Series Beds.

Figure 9. Outcrop appearance of brown sandstones within the First Flysch Beds. Below the prominent sandstone outcrops is the xenon spring.

The formations, which I mapped, are shown in figure 6. From oldest to youngest they are Chert Series Beds, First Flysch Beds, Thin Platy Limestone Beds, Thick White Limestone Beds, and Flysch Transition Beds [modified from 13 and 14]. The Chert Series Beds (Jurassic to Upper Cretaceous, up to ~100m thick) consist of thin-bedded multicolored radiolarian ribbon cherts alternating with fine-grained sandstones (figure 8). The First Flysch Beds (Upper Cretaceous, up to ~155m thick) consist of brown sandstone with interbeds of red marl, marly limestones, and limestones (figure 9). Thin Platy Limestone Beds (Upper Cretaceous, ~170m thick) are composed of even-bedded, platy limestone, and abundant chert layers and nodules (figure 10). Thick White Limestone Beds (Upper Cretaceous, up to ~300m thick) are composed of bright white limestone beds that tend to range from 1-4m in thickness, distinctively thicker than that of the Thin Platy Limestone Beds (figure 11). Flysch Transition Beds (Paleocene to Eocene, ~100m thick) constitute the uppermost unit of the Pindos Group stratigraphy and is composed of alternating thin platy to bedded limestones with layers and nodules of chert, and marly limestones (figure 12).

Figure 10. Outcrop appearance of thin bedded limestones (and concordant white chert layers and nodules) within Thin Platy Limestone Beds. Note closely spaced jointing oriented perpendicular to bedding.

Figure 11A. Outcrop appearance of Thick White Limestone Beds, which commonly is cliff forming

Figure 11B. Folding within the Thick White Limestone Beds.

Figure 12. Outcrop appearance of folded limestone within the Flysch Transition Beds.

3. Structural geology

Mt. Lykaion's dome-like summit is a thrust klippe (figure 13) containing the "Lykaion nappe," which was tectonically transported on a major thrust fault, the "Lykaion thrust". The map expression of the Lykaion thrust is a subhorizontal fault trace, which encircles most of the mountain unbroken (see figure 13). Beneath the thrust is a lower thrust plate, the "Hippodrome nappe." Archaeological elements of the upper part of the sanctuary occupy the Lykaion klippe; those in the lower part of the sanctuary occupy the Hippodrome nappe.

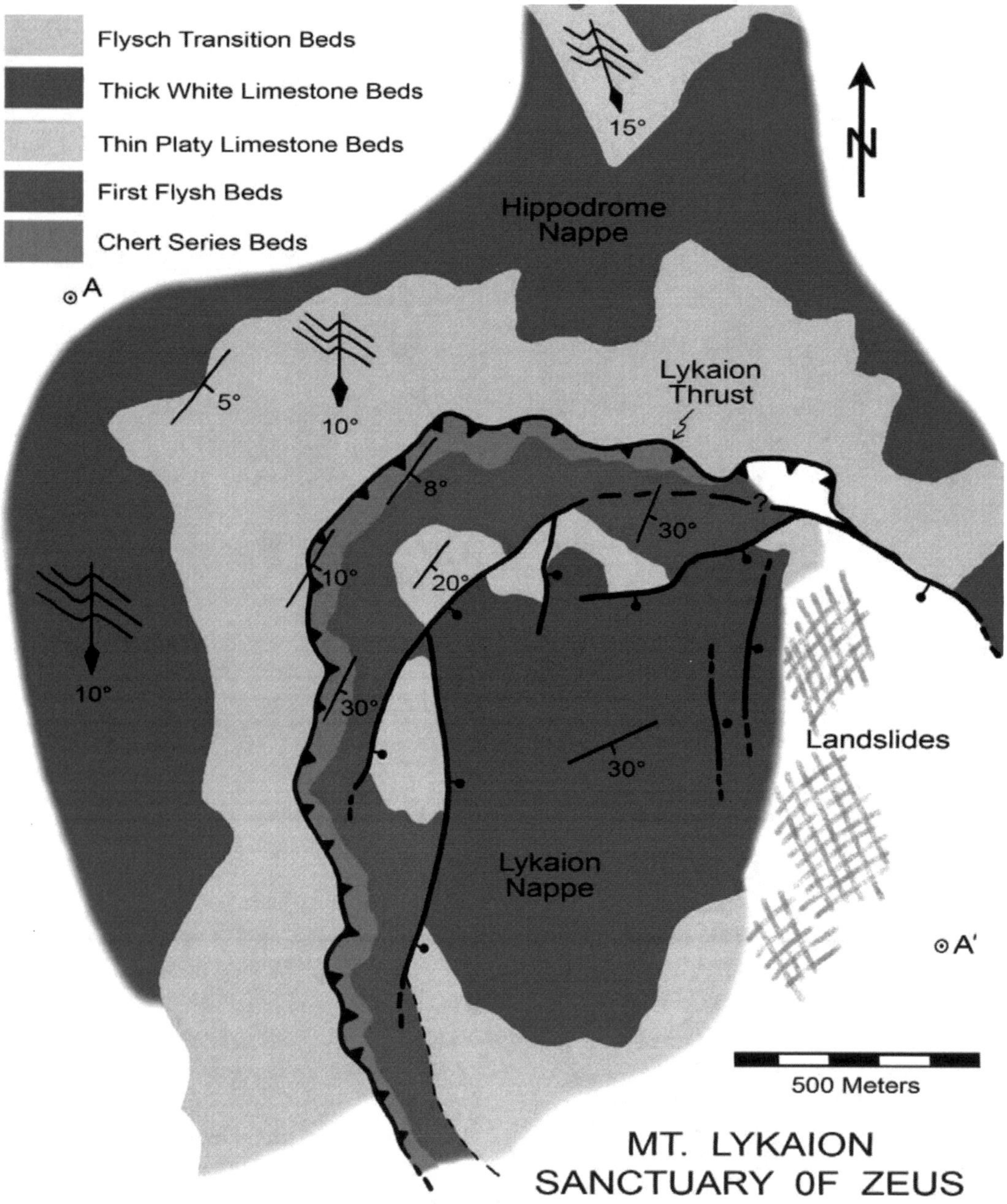

Figure 13. Simplified presentation of the geology as mapped by George Davis.

The fundamental structural relations are quite evident in the geologic map I prepared (see figure 13). The Lykaion thrust fault strikes NNE and dips ~10° SSE. Its trace can be followed from south to north along the west margin of the Lykaion klippe, where then it turns easterly along the north margin of the klippe, and (further east) proceeds to trend southeastward toward the eastern margin of the klippe. The "stacking" of thrust sheets is captured in the geologic map, and is depicted as well in a structure column I constructed to diagrammatically display key patterns and relationships (figure 14). Within both the Lykaion and Hippodrome nappes, the stratigraphic formations strike NNE and dip gently ESE.

Figure 14. Structure column depicting the structural and stratigraphic relations within the Sanctuary of Zeus at Mt. Lykaion.

4. Active tectonics

As has been well established, the Aegean region today, including Greece, is being "stretched" tectonically, a fact that is obvious in seismicity, active normal faulting, and the subsidence and tilting of the land surface [1, 15, 16, 17, 18, 19, 20, 21, 22, 23]. Global Positioning System (GPS) investigations in central Greece record velocity vectors of ~20 mm/yr, whereas velocity vectors in the southern Peloponnesus record *twice* that rate [24, 25]. This dichotomy helps us understand why the lithosphere of Greece itself, and the Aegean more broadly, is continuously stretching NE/SW. The

GPS record demonstrates that the Sanctuary of Zeus at Mt. Lykaion is moving ~37 mm/yr SW relative to Europe. This general magnitude and direction of relative movement has been occurring at least for several million years.

There are contemporary reminders of this active crustal stretching, and the power it may have imparted to the Sanctuary of Zeus at Mt. Lykaion in ancient times. One such reminder was the devastating Megalopolis earthquake (April, 1965; M5.9), the epicenter of which was merely 4 km away and 10 km below the sanctuary [1, 26, 27]. The Megalopolis earthquake was felt over an area of 300,000 km^2 (see figure 1). Its devastating effect included the killing of 20 people, the injuring of 160 more, and the collapsing of 110 houses [26]. The disturbance of unconsolidated landslide deposits, on which so many Arcadian villages are built, accounted for most of the destruction. Indeed, the village of, Ano Karyes, our host village, was *heavily* damaged by the Megalopolis earthquake.

In fact, during the course of my geological mapping I have discovered large active normal faults on the eastern edge of Mt. Lykaion, directly above the village of Ana Karyes (figure 15). Rock avalanches, rock falls, and the accelerated erosion along the active faults are the fundamental sources for the landslide complex on which Ana Karyes has been built (figure 16). The landscape features associated with this active faulting match exactly the fundamental geomorphic characteristics of active extensional faulting in Greece, as outlined by [17, 28, 29, 30]. The 'plateau' summit of Mt. Lykaion has been dropped down in a series of steps and structural terraces by the active normal faulting, which trends N-S along the eastern margin of the sanctuary.

Figure 15. SSW-directed view of Mt. Lykaion. Limestone in foreground is Thick White Limestone Beds. Road essentially follows the trace of the Lykaion thrust. Prominent limestones on middle and left skyline belong to Thick White Limestone Beds (thrust repetition with respect to foreground limestones). Far upper left captures the area of active faulting and associated scree slopes. High summit in middle background is surmounted by ash altar.

Figure 16. SSE-directed view of scarp(s) created by active normal faulting along eastern margin of the Sanctuary of Zeus, Mt. Lykaion. Bedrock is Thick White Limestone Beds.

5. Geoarchaeology

I have constructed what I call a "*geoarchaeological column*" to capture the positioning and interrelationships between bedrock, structure, and archaeology of the Sanctuary of Zeus, Mt.Lykaion (figure 17). It displays the archaeological elements in relation to the Lykaion thrust, the Lykaion and Hippodrome nappes, the stratigraphic formations within and below the Lykaion klippe, and the active normal faults. Figure 18 is a "*geoarchaeological map*" showing the archaeological elements in relation to the mapped geology.

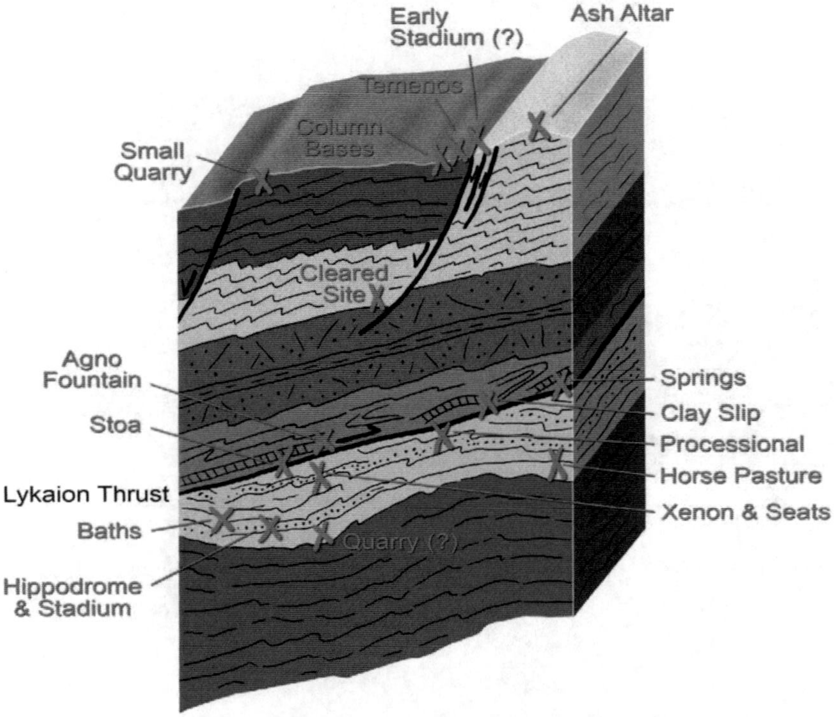

Figure 17. "Geoarchaeological Column" created by George Davis, showing archaeological elements in relationship to the stratigraphy and structure of the Sanctuary of Zeus, Mt. Lykaion.

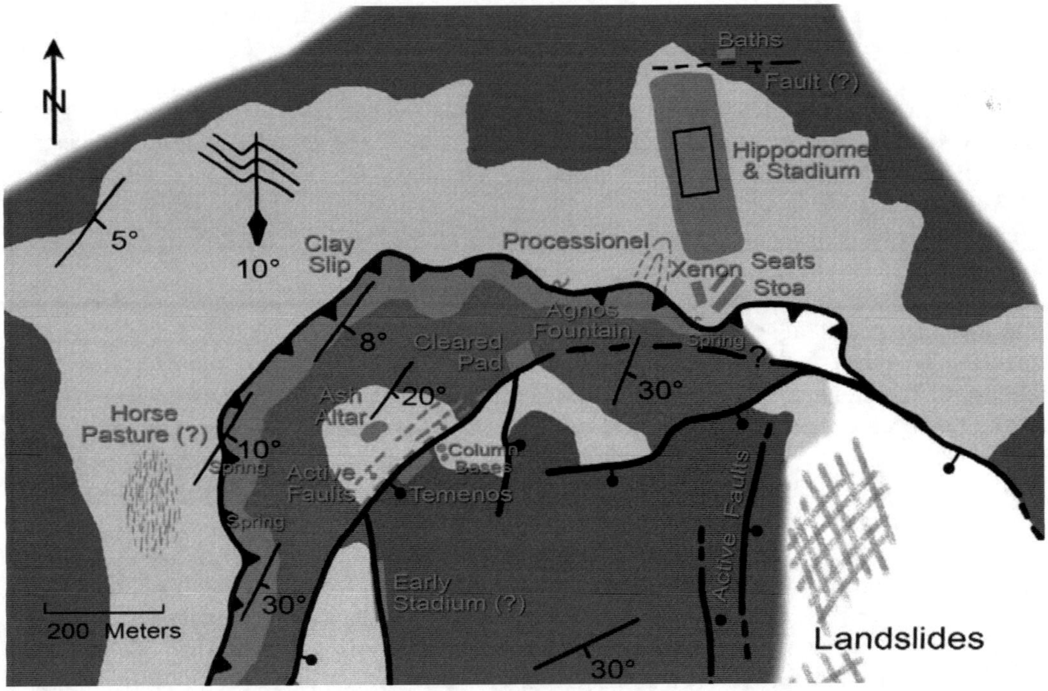

Figure 18. "Geoarchaeological Map" created by George Davis, showing archaeological elements in relationship to the mapped geology of the Sanctuary of Zeus, Mt. Lykaion.

Figure 19. The hippodrome, 300m long, occupies the flat middle ground of this south-directed photograph. The road in distance follows the Lykaion thrust. The ash altar lies at the top of the summit in upper right.

The hippodrome, which measures 300m by 125m [3] rests on Flysch Transition Beds at a stratigraphic location less than 10 to 20m above the contact with underlying Thick White Limestone Beds (figure 19). Furthermore, at this location the bedding is essentially horizontal. Differential erosion of the alternating weak and strong strata within the Flysch Transition Beds created a stripped structural surface ideal for positioning of the hippodrome. The horse pasturing area (?) occupies an identical structural and stratigraphic position (see figures 17 and 18).

The baths are built of blocks taken from nearby outcrops of Thick White Limestone Beds. The baths are in the same structural and stratigraphic location as that of the Hippodrome, but closer to the actual formation boundary between the Transitional Flysch Beds and the underlying Thick White Limestone Beds immediately to the north of the baths. The Thick White Limestone Beds dip moderately (~30°) toward the baths, and then abruptly become horizontal, as if there may be an E-W fault structure passing very close to the bath itself, thus separating two distinct bedding-dip domains (see figures 17 and 18). Perhaps the sources of water for the baths may have included artesian waters migrating upwards along such a fault zone.

The position of the stoa, which is 67m in length [3], straddles the Lykaion thrust surface (see figures 17 and 18). The foundation for the stoa is very gently dipping limestone of the Transitional Flysch Beds. Right above the foundation of the stoa are the heavily sheared and crushed Chert Series Beds, which occupy the immediate hanging wall of the Lykaion thrust. The 'trashed' Chert Series Beds would have been easily swept away to make room for the porticos of the stoa (figure 20). Less than 20m below the Lykaion thrust and positioned on Flysch Transition Beds are the seating area (39m long) and the xenon. The bedding foundation for these is very gently dipping. The building material of choice for the stoa, seating complex, and xenon is the Thick White Limestone Beds, which is also the stone of choice at the Temple of Apollo at Bassai, only ~5km from Lykaion.

Figure 20. To the left of the road are the remains of the stoa, with worked blocks from Thick White Limestone Beds. The road and most of the goats occupy the hanging wall of the Lykaion thrust. Bedrock (purple) in the immediate hanging wall is Chert Series Beds.

Springs emerge at a number of locations and always along the trace of the Lykaion thrust fault (see figures 17 and 18), which is easy to understand. Rain falling on the Lykaion klippe enters bedrock (after percolating through soil and colluvium) and descends readily through dissolution cracks and cavities within the Thick White Limestone Beds, and through joints and fractures in the Thin Platy Limestone Beds and First Flysch Beds (see figure 9). Further descent is blocked by the highly sheared, largely impermeable Chert Series Beds occupying the Lykaion thrust zone. Groundwater therefore migrates laterally and reaches 'daylight' in the form of springs along the Lykaion thrust. Two of the springs lie directly above the horse pasturing area. The stoa fountain and Agno Fountain (which Pausanias described as flowing like the Danube) lie immediately above the stoa, seating complex, and xenon.

The primary archaeological elements in the upper sanctuary are the early stadium (?), temenos, column bases, and ash altar (see figures 17 and 18). David Romano (personal communication, 2005) believes that a certain flat strip near the temenos and ash altar may have been used as a stadium at one time. The location presents itself as a narrow rectangular grassy strip measuring ~100mx~15m and trending N-S right along an active (?) normal fault contact between Thin Platy Limestone Beds on the west and Thick White Limestone Beds on the east (figure 21). There is no bedrock exposed along this strip. Furthermore, there are no blocks of rock lying on the surface within the strip, as if at one time this 'field' had been cleared.

Figure 21. SW-directed view of grassy strip, which may have been the location of an early stadium. The location of this feature is right along the trace of an active (?) normal fault that separates Thin Platy Limestone Beds on the west (right) from Thick White Limestone Beds on the east (left).

The temenos is today a flat expanse measuring ~60mx~20m, underlain by Thick White Limestone Beds (figure 22). The temenos is aligned NE-SW, and its NW border is an active (?) normal fault separating Thick White Limestone Beds on the SE from Thin Platy Limestone Beds on the NW (see figures 17 and 18). The stratigraphic level of the foundation is near the base of the Thick White Limestone Beds, where there is an interbedded limestone and mudstone transitional interval downward into the Thin Platy Limestone Beds. Bedding is essentially horizontal, and thus the flat temenos area is a stripped structural surface. Heading upslope from the temenos toward the ash altar, one immediately enters Thin Platy Limestone Beds upon crossing an active (?) normal fault. Thin Platy Limestone Beds are the underlying bedrock of the ash altar, and because this formation breaks up so readily (along closely spaced joints and thin bedding) into "cinder-block size" rocks, the ash altar hill is smooth and rounded, and lacking bench- or cliff-like limestone outcrops. Northeastward and slightly down slope from the temenos, one reaches the column bases (figure 23), which rest on Thin Platy Limestone Beds but which are composed of stones from the Thick White Limestone Beds.

Figure 22. NNW-view of temenos, with ash altar hill in background.

Figure 23. Column bases located just east of the temenos.

On the ash altar hill proper, expressions of the tendency for the Thin Platy Limestone Beds to break up into blocks are the scree deposits that occur in bands along the hill slope (figure 24). These scree bands are traces of active normal faults! The scarps produced at the times of earthquakes and ground ruptures lack the structural integrity to stand. Instead, over relatively short periods of time, the scarps (which may originally have dipped ~60°-70°) progressively collapsed to the natural angle of repose, i.e., 30°. Active faulting just a 'stone's throw' away from the ash altar much have made the sanctuary extremely powerful, especially at times of ground shaking and ground rupture.

Figure 24. NE directed view of hill surmounted by ash altar. Bedrock of this hill is Thin Platy Limestone Beds. Scree bands are traces of active normal faults. Footwall rock (Thin Platy Limestone Beds) in the scarp(s) has collapsed, thus covering the fault surface itself. Hanging wall (right) is Thick White Limestone Beds.

6. Conclusions

Tectonic factors and local geological factors both appear to have had a striking influence on the establishment of Mt. Lykaion as a sanctuary. Some of the tectonic factors include the high elevation produced by crustal shortening and thickening during the Late Cretaceous to Oligocene time, when the Pindos fold-and-thrust belt was constructed; the impressive topographic relief that is attributable to crustal stretching, and the influence of same on erosion; the circular to elliptical patterns of bedrock that are controlled by the presence of the Lykaion klippe; and the contemporary earthquake activity and active fault scarps.

Some of the local geologic factors include the suitability of Thick White Limestone Beds for blocks for the built structures; the weathering characteristics of the Thin Platy Limestone Beds, imparting a smooth conical form to the very top of the mountain; the physical and geometric properties, and the overall trace location of the Lykaion thrust; the presence of several relatively large expanses of resistant horizontal bedrock, creating flat spaces in otherwise steep mountainous country; and the geologically controlled spring systems and the relation of these to activities and trail networks.

Acknowledgements
I am very grateful to David Romano and Mary Voyatzis for their leadership in the Mt. Lykaion Survey and Excavation; and especially for assuring that investigations of the Sanctuary of Zeus would be profoundly interdisciplinary. I thank Tom Fenn, who was my field assistant during two-week stints of geological mapping in the summers of 2004 and 2005. I also acknowledge the contributions of Randy Goosen, who assisted me during a month of fieldwork in the summer of 2007. My work would not have been possible without the overall support of the Fifth Ephorate of Prehistoric and Classical Studies, whose leadership at the time of most of my work to date was Anastasia Panagiotopoulou, Ephor. Finally, I am indebted to Elizabeth Catlos for her editorial and layout support for this contribution.

7. References

[1] Ambraseys N N and Jackson J A 1990 *Int. Geophys. J.* **101** 663–708
[2] Frazer J G 1913 *Pausanias's Description of Greece* vol 4 (London) p 381–6
[3] Romano D G 1997 *Am. J. Archaeol.* **101** 374
[4] Kourouniotes K 1904 "*Αυασκαφαί Αυκαίου*" ArchEph, col. 153–214
[5] Bosellini A and Winterer E 1975 *Geology* **5** 279–82
[6] Baltuck M 1982 *Sediment. Geol.* **31** 63–88
[7] Walgreich M, Pavlopoulos A, Faupl P and Migiras G 1996 *Geol. Mag.* **133** 325–31
[8] Degnan P J and Robertson A H F 1998 *Sediment. Geol.* **117** 33–70
[9] Degnan P J and Robertson A H F 2006 *Geol. Soc. Special Pub.* **260** 467–91
[10] Avramidis P and Zelilidis A 2001 *Episodes* **24** 252–6
[11] Skourlis K and Doutsos T 2003 *Int. J. Earth Sci.* **92** 891–903
[12] Walgreich M, Pavlopoulos A, Faupl P and Migiras G 1996 *Geol. Mag.* **133** 325–31
[13] Lalechos N 1973 Geological map of the Greece Kato Fighalia Sheet (1:50,000): National Institute of Geological and Mining Research
[14] Papadopoulos P 1997 Geological Map of the Megalopolis Sheet (1:50,000; 40 m contour interval): Institute of Geology and Mineral Exploration
[15] McKenzie D 1970 *Nature* **226** 239–43
[16] McKenzie D 1978 *Geophys. J. Roy. Astr. S.* **55** 217–54
[17] Stewart I S and Hancock P L 1988 *Basin Res.* **1** 139–54
[18] Roberts S and Jackson J 1991 *Geol. Soc. London Spec. Pub.* **148** 125–42
[19] Jackson J 1994 *Ann. Rev. Earth Planet. Sci.* **22** 239–71
[20] Roberts G P 1996 *J. Struct. Geol.* **18** 835–45
[21] Reilinger R E, McClusky S C, Oral M B, King R W, Toksoz M N, Barka A A, Kinik I, Lenk O and Sanli I 1997 *J. Geophys. Res.* **102** 9983–99
[22] Cocard M, Kahle H-G, Peter Y, Geiger A, Veis G, Felekis S, Paradissis D and Billiris H 1999 *Earth Planet. Sci. Lett.* **172** 39–47
[23] Roberts G P and Ganas A 2000 *J. Geophys. Res.* **105** 23,443–62
[24] McClusky S, Balassanian S, Barka A, Demir C, Ergintav S, Georgiev I, Gurkan O, Hamburger M, Hurst K, Kahle H-G, Kastens K, Kekelidze G, King R, Kotzev V, Lenk O, Mahmoud S,

Mishin A, Nadariya M, Ouzounis A, Paradissis D, Peter Y, Prlepin M, Reilinger R, Sanli I, Seeger H, Tealeb A, Toksoz M and Veis G 2000 *J. Geophys. Res.* **105** 5695–719
[25] Oral M B, Reilinger R E, Toksoz M N, King R W, Barka A A, Kinik I and Lenk O 1995, *EOS Trans., Am. Geophys. Union* **76**
[26] Ambraseys N N 1967 *B. Seismol. Soc. Am.* **57** 1025–46
[27] Ambraseys N and Jackson J 1998 *Int. Geophys. J.* **148** 390–406
[28] Goldsworthy M and Jackson J A 2000 *J. Geol. Soc.* **157** 967–81
[29] Goldsworthy M and Jackson J A 2001 *J. Struct. Geol.* **23** 489–506
[30] Ganas A, Pavlides S B, Sboras S, Valkaniotis S, Papaioannou S, Alexandris G A, Plessa A and Papadopoulos G A 2004 *J. Struct. Geol.* **26** 2103–18

Middle Miocene E-W tectonic horst structure of Crete through extensional detachment faults

D Papanikolaou[1,3] **and E Vassilakis**[2]
[1] Professor of Dynamics and Tectonics, School of Geology and Geoenvironment, Department of Dynamics, Tectonics and Applied Geology, National and Kapodestrian University of Athens, 15784, Greece
[2] Research Scientist, School of Geology and Geoenvironment, Department of Dynamics, Tectonics and Applied Geology, National and Kapodestrian University of Athens, 15784, Greece

E-mail: dpapan@geol.uoa.gr, evasilak@geol.uoa.gr

Abstract. Two east-west trending extensional detachment faults have been recognized in Crete, one with top-to-the-north motion of the hanging wall toward the Cretan Sea and one with top-to-the-south motion of the hanging wall toward the Libyan Sea. The east-west trending zone between these two detachment faults, which forms their common footwall, comprises a tectonic horst formed during Middle Miocene slip on the detachment faults. The detachment faults disrupt the overall tectono-stratigraphic succession of Crete and are localized along pre-existing thrust faults and along particular portions of the stratigraphic sequence, including the transition between the Permo-Triassic Tyros Beds and the base of the Upper Triassic-Eocene carbonate platform of the Tripolis nappe. By recognizing several different tectono-stratigraphic formations within what is generally termed the "phyllite-quartzite", it is possible to distinguish these extensional detachment faults from thrust faults and minor discontinuities in the sequence. The deformation history of units within Crete can be summarized as: (i) compressional deformation producing arc-parallel east-west trending south-directed thrust faults in Oligocene to Early Miocene time (ii) extensional deformation along arc-parallel, east-west trending detachment faults in Middle Miocene time, with hanging wall motion to the north and south; (iii) Late Miocene-Quaternary extensional deformation along high-angle normal and oblique normal faults that disrupt the older arc-parallel structures.

1. Introduction

The first detailed geological and tectonic data of Crete were presented by Papastamatiou [1, 2] during the beginning of geological mapping at scale 1:50,000. The general picture was quite similar to that of continental Greece and Peloponnesus with the Pindos unit nappe overlying the Tripolis unit and its metamorphic basement. During the 1970's several studies resulted distinguishing a more detailed tectono-sratigraphic succession for Crete by differentiating several nappes either at the base of the nappe pile (tectonic units of Ida – Plattenkalk, Trypali, Phyllites-Quartzites) or at the top (tectonic units of Arvi, Miamou, Vatos, Asteroussia) [3, 4, 5, 6, 7, 8]. Based on the distinction between the low-grade Permo-Triassic Tyros Beds at the base of the Tripolis Upper Triassic – Eocene carbonate platform and the underlying medium grade metamorphic rocks of the Arna unit in Peloponnesus [9] a

[3] To whom any correspondence should be addressed.

modified tectono-stratigraphic succession for the Phyllites-Quartzites unit of Crete was proposed by adding the new unit of Western Crete between the Autochthon basement and the Arna unit as well as including the Variscan Sitia metamorphic rocks within the Tyros Beds [10]. The tectonic contacts separating the above tectonic units were generally considered as thrusts.

During the 1990's - 2000's several papers have proposed the interpretation that some of the previous tectonic contacts are extensional detachments usually separating the non metamorphosed nappes from the underlying metamorphosed units [11, 12, 13]. However, in several cases the amount of throw and the overall extensional deformation across the proposed extensional detachments were not demonstrated. On the contrary, it was obvious that in some cases the tectonic surfaces interpreted as extensional detachments separated the Late Triassic base of the Tripolis carbonate platform from its immediate normal stratigraphic sequence to the Tyros Beds (Middle Triassic). Thus, it was important to examine the tectonic contacts and decipher which ones were either thrusts, or extensional detachments, or surfaces of minor 'decollement' type sliding due to disharmonic phenomena, or even more complex cases with opposite kinematic characteristics of primary thrusting followed by secondary extension. This latter case would be interpreted - from a kinematic point of view - as an initial slip direction to the south, compatible with the folding and thrusting geometry of the Hellenides, followed by a younger slip direction to the north (unless, the detachments were also directed to the south).

2. Re-interpretation of the tectonic contacts of the Cretan units: thrusts or extensional detachments?

The tectono-stratigraphic nappe pile of Crete was analyzed by the re-interpretation of the geological maps combined with new field observations. Our research resulted a new map of the tectono-stratigraphic units of Crete at the scale of 1:200,000. Special emphasis was given to the distinction of the complex term "Phyllites-Quartzites" which, in fact, has to be analyzed in several tectono-stratigaphic units. Thus, from the tectonically deeper metamorphosed carbonate rocks of the relative autochthon unit of Mani (or Metamorphosed Ionian) to the shallower carbonate rocks of the Tripolis carbonate platform there is a number of different tectono-stratigraphic formations within the schistose type formations usually named in the literature as "Phyllites-Quartzites", such as (figure 1): i) The metamorphosed flysch of the Mani unit (Oligocene age, known also as Kalavros Beds), ii) The complex unit of Western Crete, comprising a Tyros type low-grade metamorphosed volcano-sedimentary sequence of Permian to Middle Triassic age, followed by Middle to Late Triassic evaporites and Late Triassic to Early Jurassic crystalline limestones (referenced as Trypali unit), iii) Medium-grade HP/LT metamorphic rocks of the Arna unit (or Phyllites-Quartzites s.s.), with the age of this unit being presumed to be Paleozoic but its metamorphism is Late Tertiary, and iv) The Permo-Triassic volcano-sedimentary base of the Tripolis platform, consisting of phyllites, meta-sandstones, volcanic rocks and some carbonate layers, equivalent to the Tyros Beds in Peloponnesus (partly known as the Ravdoucha Beds). Some evaporites and particularly the Variscan high-grade metamorphic rocks of the Sitia unit occur within the Tyros Beds in eastern Crete. Thus, the above units belong to the External Carbonate platform of the Hellenides (Terrane H_1) [14] as Mani Autochthon represents the more external part of the platform between Paxos and Ionian units, the Western Crete nappe represents a part of Ionian unit - mainly its lower part below the pelagic Middle Jurassic – Eocene sequence [15] - and finally Arna, Tyros Beds and Tripolis platform represent the central part of the external platform as the Tyros Beds occur stratigraphically underneath the platform and Arna unit represents the pre-Alpine Paleozoic basement of this platform. The presumed unconformity between the Arna basement rocks and Tyros Beds sequence has been used as a tectonic surface of discontinuity and has localized the formation of the individual tectonic unit of the Arna nappe.

The Ethia unit (equivalent to the Pindos unit in continental Greece) is usually observed superimposed on the Tripolis unit, as several nappes with limited outcrop extent all around Crete Island that occur at the higher positions in the Cretan nappe pile (figure 2). These higher nappes are: i)

Arvi unit, comprising Late Cretaceous basalts and pelagic sediments, ii) Miamou unit, comprising a late Cretaceous wild-flysch, iii) Vatos unit, comprising a Late Paleozoic – Jurassic sedimentary sequence, iv) Asteroussia nappe, comprising high-grade metamorphic rocks and granites with a Late Cretaceous age of metamorphism, and v) the Cretan ophiolitic nappe, found at the topmost position, although sometimes ophiolite bodies are also found intercalated with the Vatos and Asteroussia nappes. The aforementioned tectonic units belong to the Pindos – Cyclades tectono-stratigraphic terrane H_2 [14], with the exception of the Asteroussia metamorphics that may represent slices of the basement rocks of the internal carbonate platform of the Pelagonian terrane (H_3).

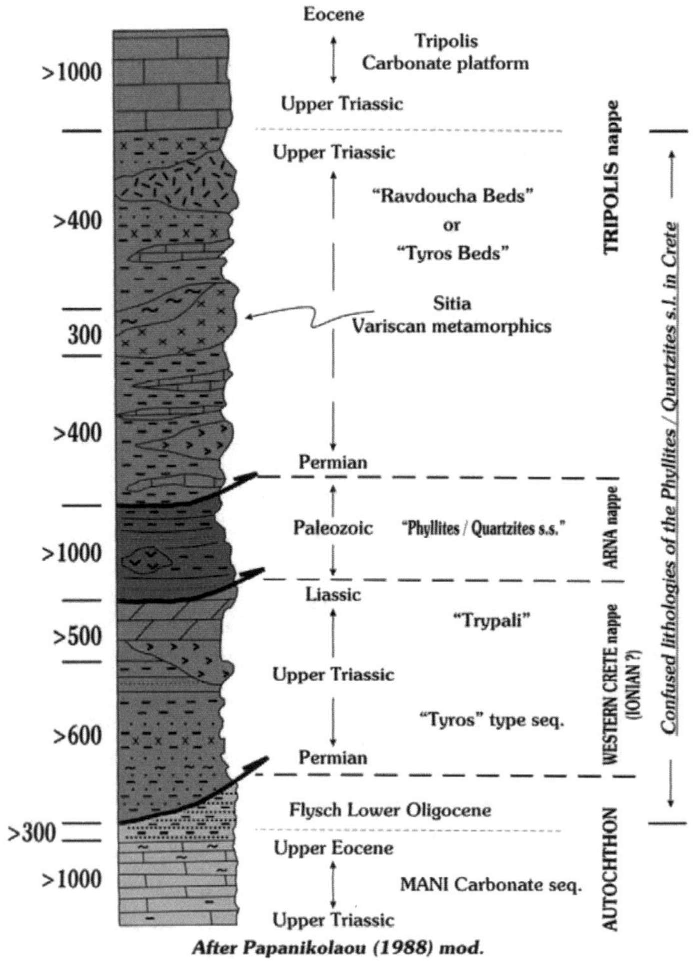

Figure 1. The most significant part of the Cretan tectono-stratigraphy which is quite often misinterpreted as Phyllites-Quartzites unit in Crete, consists of several members of four different tectonic units.

The primary tectonic contacts of the above tectonic units are thrusts, which can be observed in numerous outcrops shown on our tectonic map of Crete. The tectonic transport is from north to south, with spectacular isoclinal folds (of kilometer scale) observed in the Autochthon marbles. The overall tectonic asymmetry is comparable to that in continental Greece, where the tectonic transport is from east-northeast to west-southwest. The age of thrusting, as indicated by the syn-tectonic flysch formations of the nappes, ranges between Late Eocene in the higher nappes to Early Miocene in the lower nappes.

3. Middle Miocene Tectonic Horst Structure of Crete

Secondary tectonic contacts representing extensional detachments occur in several places but they generally form two zones of extensional detachments that strike east–west and dip 10°-30° northwards under the Cretan Sea and southwards under the Libyan sea respectively (figure 2). These two east-west trending low-angle normal faults have controlled the east-west orientation of Crete and have produced its general tectonic horst structure. However, they do not follow a particular tectonic contact but several planes of anisotropy representing both primary thrusts and stratigraphic boundaries. The structural omission across these detachments varies, in some cases the whole tectono-stratigraphic column of Crete is missing, such as in the area of the southern slopes of Psiloritis Mt, where ophiolites of the uppermost unit are resting on the lowest unit of the autochthon marbles [16]. This structure is similar to the Parnon detachment in the Peloponnesus regime [17]. In some other places, significant parts of the Cretan nappe pile are missing, like in the area east of Viannos, where the upper nappes of Arvi and Pindos occur at the hanging wall next to the Tripolis carbonates which are located at the footwall. The Psiloritis and Viannos detachments belong to the southern detachment of Crete together with the detachment observed along the southwestern coastline of Crete in Cape Cryos, Paleochora and Sougia. Here, relics of Tripolis and overlying Pindos nappe dip northwards against the detachment whereas the Western Crete unit forms the footwall. The northern detachment of Crete can be observed in the area of Platanos, Sfinari and Topolia in northwestern Crete, where the Pindos and Tripolis nappes occur in the hanging wall whereas Arna unit forms the footwall. Another fault belonging to the northern detachment occurs in the area of Spili and Vatos in central Crete where the Vatos unit imbricated with the ophiolites occurs in the hanging wall whereas the Arna and Tripolis units form the footwall.

The age of the extensional detachments is Middle Miocene as dated by syn-tectonic sediments of breccia-conglomerates (e.g. the Topolia formation) [18]. Some impressive sub-horizontal contacts observed between the Tripolis carbonates and the Tyros Beds (as in the area of Sitia) are minor structures without any important structural omission. The sharp contact observed in several outcrops is due to disharmonic phenomena developed during the compressive primary and/or the extensional secondary deformation.

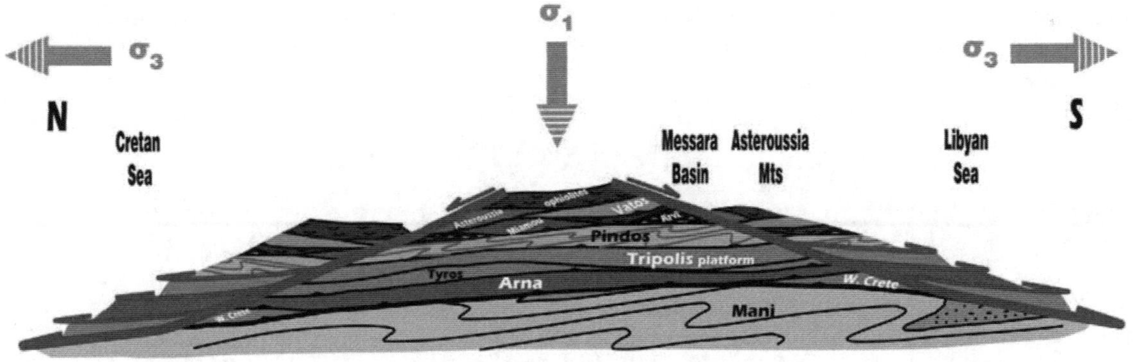

Figure 2. The Cretan nappe pile which was the result of the compressional phase later modified by arc-parallel extensional detachments dipping either northwards or southwards causing significant structural omission. The opening of the Cretan basin is related to the northern detachment and the Messara basin above the Asteroussia mountain is related to the southern detachment.

The above east-west trending extensional detachments of Crete represent arc-parallel extensional structures following the previous arc-parallel compressional thrusts and folds. The tectonic horst structure of Crete is related to the opening of the Cretan basin in the north whose southern margin is controlled by the northern detachment of Crete and western margin by the East Peloponessus detachment [17]. An opposite facing detachment has been described in Anafi Island along the northern margin of the Cretan basin with a similar succession of deformational phases [19]. The southern detachment of Crete is related to the opening of the Messara basin between Psiloritis mountain and Asteroussia mountain and its westward prolongation between southwestern Crete and Gavdos [16]. It is characteristic that all over the hanging wall of this southern detachment in Asteroussia and Gavdos the Alpine formations beneath the Middle Miocene - Quaternary sediments belong to the higher nappes (from Pindos to Asteroussia and the ophiolites).

High-angle normal and strike-slip faults dissect and disrupt the previous structure of thrusts and extensional detachments and form the tectonic margins of the Late Tortonian – Quaternary basins. The majority of such faults are seismically active structures overprinting the previous tectonic fabric. Nevertheless, some segments of the former detachments are still active today. The outcrops of the Late Miocene – Pliocene marine sediments show the blocks that subsided during this period. The general uplift of Crete during Quaternary time marks the present-day deformation with half of the island being a recently emerged area from the previously submerged neotectonic grabens.

4. Conclusions

Two east-west trending extensional detachments have been recognized in Crete with a northward slip below the Cretan Sea and a southward slip below the Libyan Sea. These two detachments bound a general east-west trending tectonic horst structure of Crete which was created during Middle Miocene time. These detachments disrupt the overall tectono-stratigraphic succession of Crete and use as sliding surfaces either the pre-existing thrusts or some stratigraphic boundaries representing strong anisotropies, such as the transition from the Permo-Triassic Tyros Beds to the base of the Late Triassic – Eocene carbonate platform of the Tripolis nappe. The distinguishing of the different tectono-stratigraphic formations occurring in what has been usually mentioned as "Phyllites-Quartzites" unit helps to decipher the real detachments from the thrusts or the minor discontinuities. Thus, the tectonic contacts developed between the Tyros Beds and the base of the Tripolis carbonate platform are not major extensional structures. In contrast, the detachment occurring along the southern slopes of Psiloritis Mt has omitted the entire Cretan tectono-stratigraphic succession (a thickness of about 10-12 km with a southward tectonic displacement of about 30-35 km). The deformational history of Crete can be summarized in three stages: (i) a first arc-parallel compressional phase with east-west trending thrusts with a southward tectonic transport during Oligocene – Early Miocene time, (ii) a second arc-parallel extensional phase of east-west trending detachments with transport both to the north and to the south during Middle Miocene time, and (iii) a third extensional phase of higher angle normal and oblique normal faults which disrupt the previous arc-parallel structures during Late Miocene – Quaternary time.

Acknowledgements
We thank B.C. Burchfiel (MIT) for his comments and reviewing our manuscript.

5. References

[1] Papastamatiou J, Vetoulis D, Bornovas J, Christodoulou G and Tataris A 1959 Geologische Karte von Kreta (1:50.000) Blatt Sitia. Inst. Geol. Subsurface Res., Athens
[2] Papastamatiou J et al 1959 Geologische Karte von Kreta (1:50.000) Blatt Ierapetra. Inst. Geol. Subsurface. Res., Athens
[3] Bonneau M 1984 *Geol. Soc. London Spec. Pub.* **17** 517–27
[4] Bonneau M, Angelier J and Epting M 1977 *Bull. Soc. Geol. France* **7** 87–102
[5] Creutzburg N et al. 1977 General geological map of Crete (scale 1:200.000) I.G.M.E., Athens
[6] Krahl J et al 1983 *Geol. Rundsch.* **72** 1147–66
[7] Seidel E, Okrusch M, Kreuzer H, Raschka H and Harre W 1976 *Contrib. Mineral. Petrol.* **57** 259–75
[8] Wachendorf H, Best G and Gwosdz W 1975 *Geol. Rundschau* **64** 728–50
[9] Papanikolaou D and Skarpelis N 1987 *Ann. Geol. Pays Hellen.* **33** 47–68
[10] Papanikolaou D 1988 *Introduction to the geology of Crete.* IGCP Project No 276, 1988 Field

meeting in Chania, Field guide 36 pp.
[11] Fassoulas C, Kilias A and Mountrakis D 1994 *Tectonics* **13** 127–38
[12] Jolivet L et al. 1996 *Tectonics* **15** 1129–53
[13] Ring U, Brachert T and Fassoulas C 2001 *Terra Nova* **13** 297–304
[14] Papanikolaou D 1997 *Ann. Geol. Pays Hellen.* **37** 495–514
[15] Papanikolaou D 1979 Stratigraphy and structure of the Palezoic rocks in Greece: An introduction. *Correlation of Prevariscan and Variscan events of the Alpine-Mediterranean mountain belt* ed F P Sassi IGCP Project No.5 Newsletter **1** 93–102
[16] Vassilakis E 2006 *Study of the tectonic structure of Messara basin, central Crete, with the aid of remote sensing techniques and G.I.S.* PhD Thesis (Athens, Greece: N K University of Athens) 546 pp
[17] Papanikolaou D and Royden L 2007 *Tectonics* **26** TC5003 doi:10.1029/2006TC002007
[18] Meulenkamp J E 1979 Field guide to the Neogene of Crete. Publications of the Department of Geology and Paleontology, Series A, No 32. Univ. Athens, 32 pp.
[19] Soukis K and Papanikolaou D 2004 *Bull. Geol. Soc. Greece* **XXXVI** 1688–96

IOP Conf. Series: Earth and Environmental Science **2** (2008) 012004 doi:10.1088/1755-1307/2/1/012004

Dating movement in shear zones: The example of the South Cyclades Shear Zone, Ios, Aegean Sea, Greece

Marnie Forster[1] **and Gordon Lister**
Research School of Earth Sciences, The Australian National University, Canberra 0200, Australia

E-mail: marnie.forster@anu.edu.au

Abstract. The South Cyclades Shear Zone (SCSZ) is a ~1 km thick bowed-up ductile shear zone that defines the carapace of a deeply eroded gneiss dome that outcrops on the island of Ios in the Cycladic archipelago, Aegean Sea, Greece. The discovery of the SCSZ played a pivotal role in the recognition of the Aegean metamorphic core complexes, and this supposedly Miocene shear zone has been studied in detail by several authors. This contribution illustrates the difficulties involved in systematically determining the age of deformation fabrics using $^{40}Ar/^{39}Ar$ geochronology, and provides a case study that shows the value of carefully designed step-heating experiments on microstructurally controlled samples. In samples such as these, from complex metamorphic tectonites, laser fusion, or step heating experiments with relatively few steps, would have led to erroneous conclusions.

1. Scope of work

This presentation to the Harrington symposium is based on a paper currently in review [1]. Therein we report new $^{40}Ar/^{39}Ar$ geochronology from the South Cyclades Shear Zone (SCSZ), Ios, Cyclades, Greece. This contribution builds on the material submitted in that paper, specifically addressing the link between $^{40}Ar/^{39}Ar$ geochronology and Rb-Sr geochronology. In this contribution we explicitly exclude discussion of results from *in situ* $^{40}Ar/^{39}Ar$ geochronology using laser ablation, since this method is inherently limited in a complex microstructure. Laser ablation in principle produces mixing ages, and only under exceptional circumstances is the method able to discriminate the isotopic characteristics of end member microstructural reservoirs. The same limitations do not apply to $^{40}Ar/^{39}Ar$ geochronology using carefully designed step-heating experiments, in particular if the apparent age spectra that result can be analyzed using the method of asymptotes and limits [2]. The recognition of frequently measured ages (or FMAs) proves critical in this aspect, and must be considered in addition to the points noted in this paper.

The essential message we bring is that there is not a single "age" that should be expected as the result of the application of step heating $^{40}Ar/^{39}Ar$ geochronology. A metamorphic tectonite potentially is capable of recording the effects of a considerable extent of its history, and techniques that purport to record a single "age" in respect to this history are thus inherently flawed in respect to the logic that has been applied. This blunt assessment is valid except in cases where such a measured "age" can be

[1] To whom any correspondence should be addressed.

associated with a single well-defined short-lived event. Random guesses as to the significance of misconstrued 'plateau' do not qualify. Nor does an arbitrarily determined average age offer a valid application of $^{40}Ar/^{39}Ar$ geochronology in this context. The science of dating microstructure in a complex deformed and metamorphosed tectonite has progressed beyond such a point.

The SCSZ is a ~1 km thick bowed-up ductile shear zone that defines the carapace of a deeply eroded gneiss dome that outcrops on the island of Ios in the Cycladic archipelago, Aegean Sea, Greece. The discovery of the SCSZ played a pivotal role in the recognition of the Aegean metamorphic core complexes, and this supposedly Miocene shear zone has been studied in detail by several authors. We had returned to the SCSZ to further our research into new methods for argon geospeedometry, and to test the shear heating hypothesis. Earlier work [3,4] led to the conclusion that short thermal pulses had been associated with the operation of this extensional ductile shear zone. In the process of this reanalysis, however, we discovered that the published "age" for this shear zone was incorrect. In addition we discovered that K-feldspar is considerably more retentive than previous analyses would suggest, allowing its use to directly date mylonitic deformation that cataclases large pre-existing phenocrysts. This contribution thus makes an advance in reassessing methodologies currently in use for "dating" shear zones, as well as furthering our understanding of Aegean geology.

The SCSZ is a key structural element in the tectonic evolution of the Cycladic massif because it juxtaposes the Cycladic Eclogite Blueschist Unit (EBU) above, against a Hercynian magmatic and metamorphic basement complex, below. Our new, microstructurally focussed $^{40}Ar/^{39}Ar$ geochronology, suggests the SCSZ initiated just before the time of the Eocene-Oligocene transition (at ~35 Ma) [1]. Therefore it is not only Miocene extension that led to the Tethyan metamorphic core complexes exposed in the Cycladic archipelago. It is shown here that pervasive top-to-the-south directed ductile shear took place in the SCSZ for ~5 Ma during the early Oligocene, followed by initiation of overprinting north-sense shear zones at the time of the Oligocene-Miocene transition (at ~25 Ma) [1]. These results mean that modern (core complex related) extension in the Aegean started much earlier than has been previously recognized, and that stretching of the Aegean lithosphere took place in a distinctly episodic fashion, not in a manner that can be described as continuous, or progressive. The data also mean that the complementary shear-sense geometry of the SCSZ and the North Cyclades Shear Zone (NCSZ) cannot be interpreted as crustal-scale boudinage initiated during crustal extension in Miocene time, since the opposing sense shear zones operated at distinctly different times. Similarly, models that suppose exhumation of the Cycladic Eclogite-Blueschist Unit as the result of extrusion can also be eliminated. The research involved new step heating experiments using irradiated white mica and K-feldspar. The method of Tectonic Sequence Diagrams (or TSDs) was applied [5] so as to allow microstructural definition of the gas reservoirs being sampled. The TSDs show that the microstructure reflects the imprint of a succession of several distinct thermal events.

We begin by showing how K-feldspar can be used to date mylonitization, in particular where cataclasis and recrystallization of large pre-existing phenocrysts is induced by deformation. The first point demonstrated was that the ages obtained are not 'cooling ages'. The Arrhenius data showed that the K-feldspars analyzed were considerably more retentive than the "average" UCLA K-feldspar [6]. The activation energy obtained using a multi-diffusion domain compliant analysis was up to 1.5 to 1.8 times greater than the UCLA average, and in many cases fell outside of the data spread reported from UCLA data by [6]. The reason for this substantial discrepancy is as yet unknown. Irrespective of the cause it has considerable ramifications for the use of K-feldspar in $^{40}Ar/^{39}Ar$ geochronology.

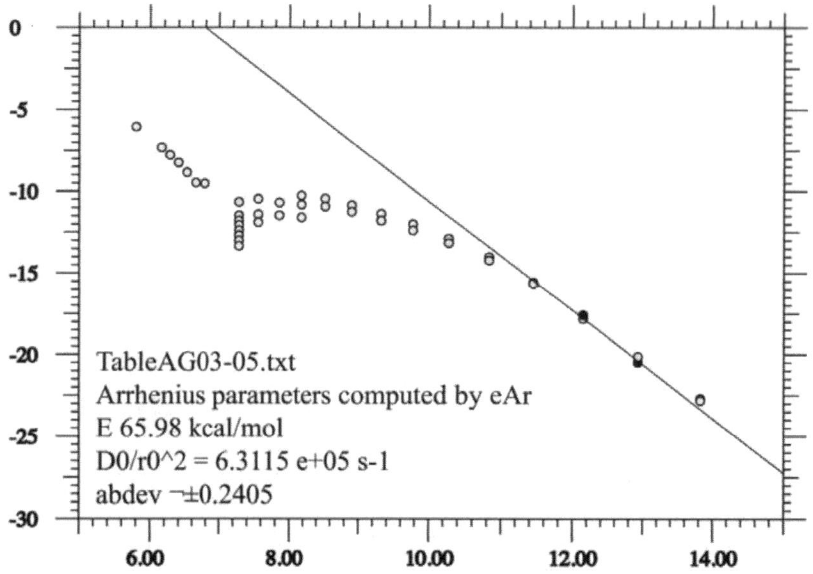

Figure 1. An Arrhenius plot produced from sample AG03-05 using Program eAr (available from http://rses.anu.edu.au/tectonics/programs/).

Sample AG03-05 was taken from a north-sense shear zone that overprints the SCSZ. This shear zone cataclased and/or recrystallized (originally Hercynian) K-feldspar porphyroclasts relict from previous events. Figure 1 shows the Arrhenius data obtained by analysis of this step heating experiment, while figure 2a shows the apparent age spectrum with the asymptotes and limits superimposed. Figure 2b shows an age probability plot derived using these asymptotes and limits. Note that the experiment involved isothermal duplicates for each temperature measured.

Because the K-feldspar was retentive it was possible to directly date cataclasis and recrystallization associated with the overprinting ~25 Ma N-sense shear zones. The activation energy is ~1.5 times greater than the UCLA average reported by [6]. This means that the Argon Partial Retention Zone for this K-feldspar encompasses P-T conditions appropriate to the operation of the SCSZ. Therefore a correlation between age and microstructure can be expected, and such a correlation was observed as predicted by this result. The least retentive domains record the effects of a thermal event related to Miocene plutonism. This is evident in the microstructure in the form of discrete zones in which sericitization of K-feldspar, takes place (but note that such sericite was not developed in the measured sample). Based on results such as those illustrated in figure 2, the age of this hydrothermal event is estimated as taking place at some time younger than ~13-14 Ma.

These results will generate debate. However, whereas in principle we agree with [7] that microstructures in K-feldspar preclude simple-minded application of multi-domain diffusion theory, to their comment we add that their assumption of a monotonically decreasing cooling path is similarly unwarranted. Even granites in apparently stable terranes can be subject to later thermal and/or fluid pulses and such activity is not uncommon during PanAfrican events in the terranes in question. With appropriate caution, microstructures produced as the result of a later imposed event can be analyzed to glean additional information as to the thermal and tectonic evolution of the rocks in question. The basement terrane on Ios is similarly subject to a sequence of thermal/fluid pulses, some of which were associated with deformation events. These events produced microstructural modifications that can be systematically analyzed to reveal information as to the timing of the individual events.

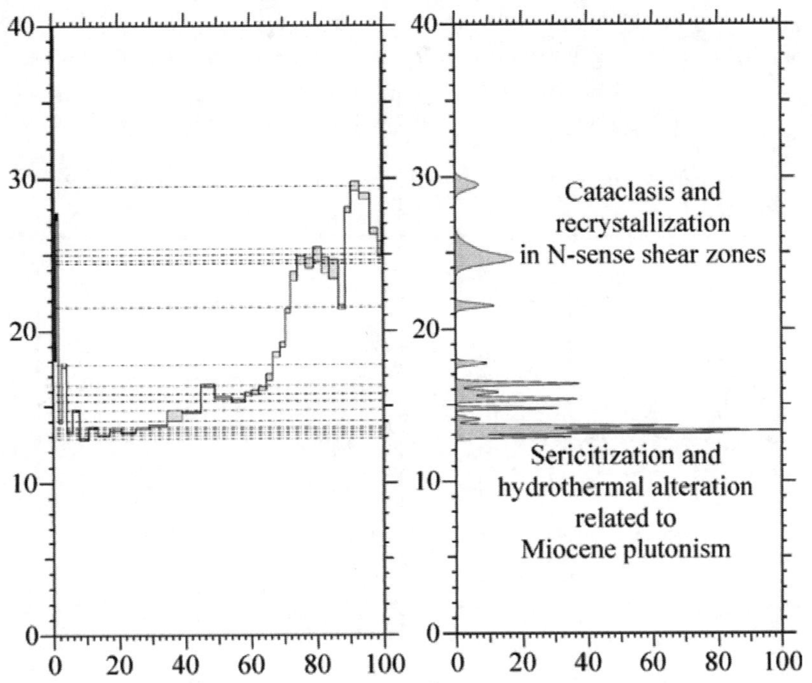

Figure 2. Apparent age spectrum (a) for a step heating experiment on sample AG03-05 showing asymptotes and limits derived using *Program eAr*. This program is freely available and can be downloaded from <http://rses.anu.edu.au/tectonics/programs/>. In (b) an age probability plot shows the maximum age for sericitization and hydrothermal alteration related to Miocene plutonism (~13-14 Ma), and the minimum age for the onset of cataclasis and recrystallization of K-feldspar in N-sense ductile shear zones (~25 Ma).

This research illustrates the difficulties involved in systematically determining the age of deformation fabrics using $^{40}Ar/^{39}Ar$ geochronology, and provides a case study that shows the value of carefully designed step-heating experiments on microstructurally controlled samples. In samples such as these, from complex metamorphic tectonites, laser fusion, or step heating experiments with relatively few steps, would have led to erroneous conclusions. This point is illustrated in figure 3, showing white mica from a similar N-sense overprinting shear zone. The shear zone has led to almost total recrystallization of relict fabrics, so these now record an age close to the time that the shear zone commenced operation. In contrast, the first steps of the age spectrum record the time that the shear zone ceased operation. The temperatures would have been relatively low by this time, so that micas were shredded rather than recrystallized. Partial argon loss would result from the commensurate decrease in effective diffusion dimension [cf 1].

Finally we discuss the interpretation of $^{40}Ar/^{39}Ar$ age spectra in the context of Rb-Sr geochronology from the same ductile shear zones. There are specific views as to the relationship between 'ages' produced by the two methodologies. For example [8] consider that $^{40}Ar/^{39}Ar$ age spectra are generally "*disturbed*" with total fusion ages "*3–6 Ma older than corresponding Rb–Sr ages*". They suggest inheritance from the high-pressure stage of the tectonic history of these rocks as the most likely cause for the complex age spectra and the older dates, and they note that this "*suggests that the Rb–Sr system is more sensitive to modification during overprinting than the Ar isotope system*". In contrast we contend that $^{40}Ar/^{39}Ar$ total fusion ages are of no relevance (figure 4). More importantly, many of the phases measured for the Rb-Sr analysis more than likely grew in the shear zone at some random time during its operation, and therefore equilibriated with the grain boundary pore fluids at that time. In contrast, relict mica can retain considerably more isotopic information, and thus provide information throughout all stages of the history of deformation and metamorphism.

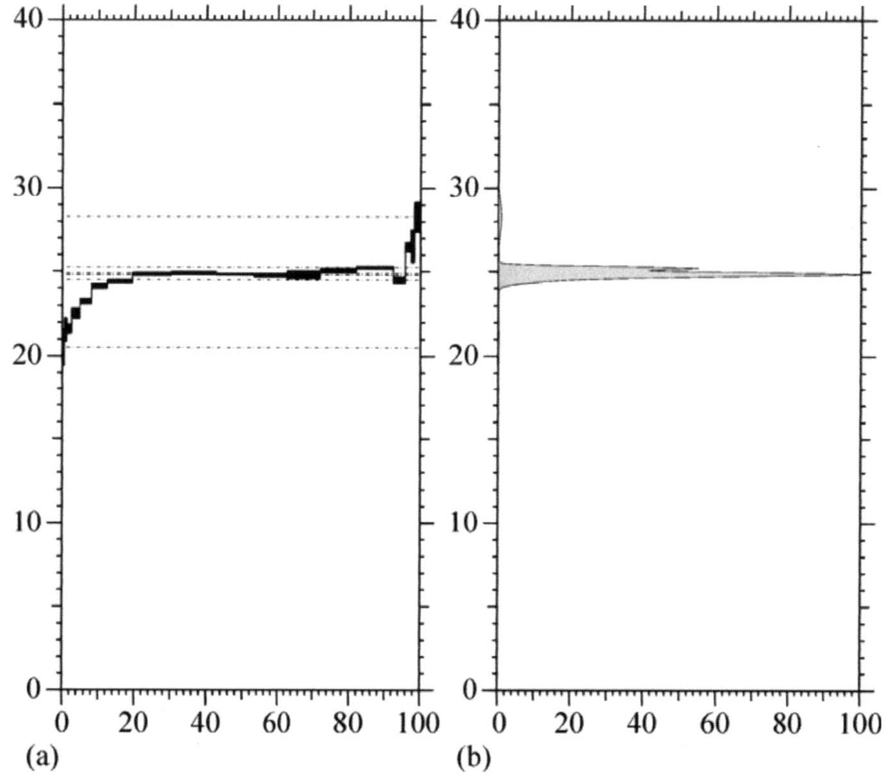

Figure 3. Apparent age spectra (a) for white mica from a similar N-sense overprinting shear zone, with asymptotes and limits superimposed. In (b) this data is used to derive an age-probability plot. The shear zone has led to almost total recrystallization of relict fabrics. The upper asymptote thus records the time that the shear zone commenced operation. The first steps of the age spectrum reflect the time that the shear zone finished its operation.

We suggest that $^{40}Ar/^{39}Ar$ age spectra measured by [8] reflect age variation in a composite fabric within the ductile shear zone. Based on data reported from one of their samples the shear zone operation would have commenced operation some time around ~25 Ma, and continued operation until some time after ~23.5 Ma (figure 4). The bulk of the age spectrum reflects the partially reset ages in the composite relict fabric, however. Our data show that the variation in age in this part of the spectrum can be considerably greater than the 3-6 million years suggested by [8].

There is thus no point in attempting to assign a single "age" to a $^{40}Ar/^{39}Ar$ apparent age spectrum, and even less value in attempting to use a bulk fusion age to systematically assess the age of operation of a shear zone. In contrast, Rb-Sr data may indeed reflect the age of minerals isotopically re-equilibriated during their fluid-assisted growth and recrystallization at some time during shear zone operation. However, because a single "age" is determined, potentially far less information is available as to the timing of the start and end of shear zone operation. In addition, because only single numbers result from an isochron, the Rb-Sr age can be a mixing age in a time range determined by different periods of fluid assisted growth and/or recrystallization in the evolving shear zone.

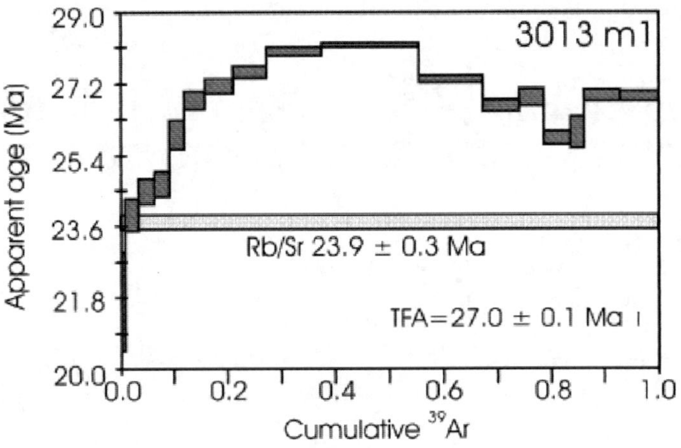

Figure 4. Part of a figure published by [8] illustrating the relationship between a complex $^{40}Ar/^{39}Ar$ apparent age spectrum and a single Rb-Sr age. Based on the results reported here the Rb-Sr age may define an age close to when a shear zone ceases operation.

In terms of dating the movement in ductile shear zones there is thus no panacea. Considerable effort will be involved. Nevertheless, carefully designed, microstructurally focussed step heating experiments using $^{40}Ar/^{39}Ar$ geochronology can be systematically applied to date the evolution of fabric in low temperature ductile shear zones such as are developed throughout the Cyclades. Laser methods are not capable of replicating such results since each fusion experiment yields a single age and are thus not capable of discriminating effects related to mixing of gas reservoirs at small scales in a microstructure.

2. References

[1] Forster M A and Lister G S 2008 *J. Geophys. Res.* in review
[2] Forster M A and Lister G S 2004 *J. Struct. Geol.* **26** 287–305
[3] Lister G S and Baldwin S L 1996 *Tectonophysics* **253** 83–109
[4] Baldwin S L and Lister G S 1998 *J. Geophys. Res.* **130** 7315–36
[5] Forster MA and Lister G S 2008 *J. Geol. Soc. London* in press
[6] Lovera O M, Grove M, Harrison T M and Mahon K I 1997 *Geochim. Cosmochim. Acta* **61** 3171–92
[7] McLaren S, Dunlap W J and Powell R 2007 *J. Geol. Soc.* **164** 941-944
[8] Bröcker M, Bieling D, Hacker B and Gans P 2004 *J. Metamorph. Geol.* **22** 427–42

Do variations in Arabian plate lithospheric structure control deformation in the Arabian-Eurasian convergence zone?

Robert J Stern[1,3] and Peter R Johnson[2]
[1]Dept. of Geosciences, University of Texas at Dallas, Richardson TX 75080-3021, USA
[2]Johnson and Vranas Associates, 1242 Tenth Street, NW, Washington, DC 20001-4214, USA

E-mail: rjstern@utdallas.edu

Abstract. The Arabian plate has been converging with Eurasia for 20-30 Ma, currently at 2-3 cm/year. Convergence is manifested differently along strike, with collision and tectonic escape in the west (Anatolia) and subduction of Arabia beneath Eurasia in the east (Iran). The reason for these differences may reflect the greater density of the Arabian lithosphere in the east relative to that in the west. Five observations indicate that the eastern Arabian plate is more dense and thus easier to subduct than the western Arabia Plate: (1) the >1000 km-long, N-S trending Central Arabian Magnetic Anomaly (CAMA) marks a ~600Ma-old suture between E and W Arabia that (2) separates crustal tracts with distinct Neoproterozoic histories (850-750 Ma crust in the east, 850-570 Ma in the west); (3) E. Arabian (platform) crust is slightly thicker and more felsic than that of W. Arabia (shield) and thus more buoyant; (4) E. Arabia is underlain by thicker mantle lithosphere than W. Arabia; and (5) W. Arabia has a long history of uplift whereas E. Arabia has subsided throughout Phanerozoic time. We infer that regional variations in lithospheric thickness largely reflect the earlier stabilization (and onset of conductive cooling) of E. Arabia relative to W. Arabia, leading to the development of significantly thicker mantle lithosphere in the east relative to the west. This also explains why Arabia east of CAMA was inundated by shallow seas ~540 Ma ago and accumulated salt whereas Arabia west of CAMA did not, as well as the long history of subsidence of the Arabian Platform and uplift of the Arabian Shield. These differences in lithospheric thickness may have been further modified by thermal erosion in Cenozoic time due to Red Sea rifting and the Afar hotspot.

1. Introduction

The Arabian Plate is mostly continental crust and upper mantle, flanked on three sides by narrow strips of oceanic lithosphere. The Arabian Plate converges with Eurasia to the north at 2-3 cm/year [1]. There has been roughly 300–500 km of convergence since Arabia first arrived at the trench 20–30 Ma ago [2], but plate convergence is partitioned differently along strike. Subduction north of the Persian Gulf forms the Zagros fold-and-thrust belt and Urumieh-Dokhtar magmatic arc, whereas continent-continent collision in the west produced the Bitlis suture, delamination and related igneous activity in eastern Anatolia, and forces the Anatolian plate to escape westward along strike-slip faults (figure 1)[3]. The reasons for this striking difference between eastern and western parts of the convergence

[3] To whom any correspondence should be addressed.

zone may result from differences between the lithospheres of E and W Arabia.

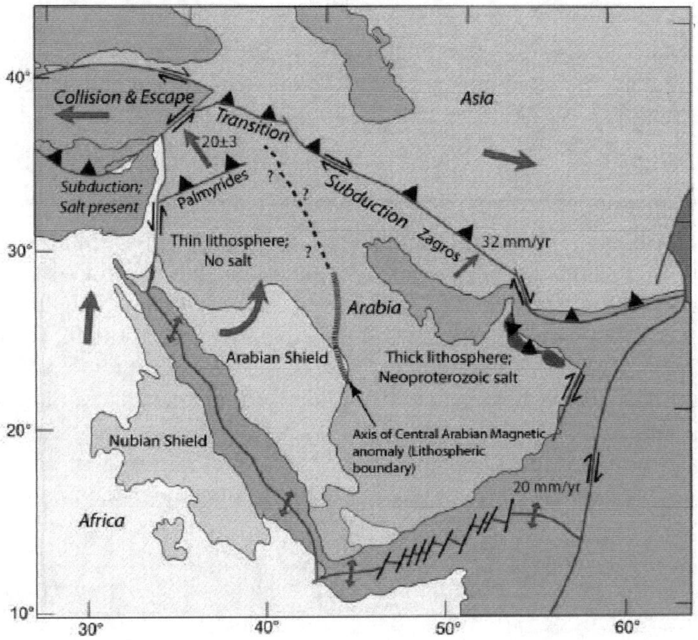

Figure 1. Simplified map of the Arabian and surrounding plates, with plate boundaries (red) and convergence vectors (red arrows). Grey areas are exposed basement. The Central Arabian Magnetic Anomaly marks a fundamental boundary between lithospheres of eastern and western Arabia; northward extension of CAMA (dashed with question marks) is speculative.

2. Formation of Arabian Plate Continental Lithosphere

Upper continental crust exposures dominate western Arabia (Arabian Shield) and occur as scattered small outcrops in eastern Arabia. Basement rocks mostly yield Neoproterozoic radiometric ages (but with significant tracts of Paleoproterozoic and Archean crust in the southern Shield and Yemen). Geochronologic and isotopic data indicate that most Arabian plate crust formed during about 300 Ma of Neoproterozoic time, 870 to 570 Ma ago. Arabian Shield crust makes up about half of the Arabian-Nubian Shield (ANS). The Arabian-Nubian Shield comprises one of the great tracts of juvenile continental crust on Earth and the northern half of the East Africa-Antarctic Orogen (EAAO) [4], which formed when large fragments of Eastern and Western Gondwana and intervening oceanic arcs collided ~620-570 Ma ago. Collisional tectonics to form crust of the Arabian Shield was succeeded by extension/collapse resulting in inundation, as manifested by deposition of carbonate sediments like those of the ~590 Ma Muraykhah Formation [5] and Dhaiqa Formation [6]. The nature of buried crust east and north of the Arabian Shield is less clear. The Central Arabian Magnetic Anomaly (CAMA) trends N-S for 800 km and can be traced due north across the Saudi Arabia-Iraq border, perhaps as far as the Zagros Orogenic Belt [7]. CAMA can be traced onto the easternmost Shield as the Al Amar fault, also known as the Amar suture [8]. This suture separates igneous rocks of the >689-625 Ma Ar Rayn magmatic arc [9] to the east from granites and thick sediments (Abt formation) of the Ad Dawadimi terrane to the west. CAMA thus marks a fundamental tectonic boundary that formed ~600 Ma ago [10], between juvenile Neoproterozoic (ANS-like) crust to the west and poorly known crust to the east. Crust to the east of CAMA is poorly known. Scattered basement exposures east of CAMA are Neoproterozoic, with no evidence for pre-Neoproterozoic ages or isotopic compositions. Our present understanding of E. Arabia crust is that igneous activity ended and the crust began to stabilize ~750 Ma ago, almost 200 Myr before stabilization of W. Arabia, as recorded in rocks of the Arabian Shield [11, 12, 13, 14, 15, 16].

One of the most important late Neoproterozoic events affecting the western part of the Arabian Shield was development of NW-SE trending Najd faults and shear zones during collision between Eastern and Western Gondwana. Displacements as large as 400 km are inferred across some of these faults [17], which occurred ~580-640 Ma ago. These shears are thought to have resulted as far-field effects due to continent-continent collision to form the southern EAAO [18]. These shears are steep at the surface but likely flatten at depth, reflecting the fact that Arabian crust was still quite hot at this time, and mid-crustal levels were almost certainly weak. Such listric faults, especially where these interacted with the abundant intrusions, were surely instrumental in shaping the structure of the crust that we observe today. It is not clear whether this shear system affected eastern Arabia; a magnetic map of central Arabia shows the N-S trending CAMA as continuous from 21° to 29°N, with no indication of significant sinistral offsets [8].

3. Phanerozoic Sedimentation and Subsidence

Clues to lithospheric structure may be found in the thick sedimentary cover of the Arabian Platform, which contrasts with uplifted crust of western Arabia. One possibility to explain the Shield/Platform dichotomy is that the Shield was dynamically uplifted recently due to hot mantle of the Afar plume and because of Red Sea rifting, whereas the Platform was not. Another possibility is that much older crust and lithosphere underlies the Arabian Platform whereas more buoyant crust and lithosphere of Neoproterozoic age underlies the Arabian Shield. A third possibility is that crust beneath the Shield and Platform are of similar age but are associated with a lithosphere is significantly thicker beneath the Platform. These different explanations imply different timings for the structural differentiation of Arabian plate crust into shield and platform. Choosing the best explanation requires understanding the depositional history of Arabian Plate sediments, which preserve the history of regional uplift and subsidence.

Regions of Arabia now defined as Platform and Shield behaved similarly until about ~750 Ma ago, with tectonics and sedimentation dominated by active magmatic activity across a broad region; these blocks coalesced by ~560 Ma. But after ~750 Ma ago, eastern Arabia stabilized as a neocraton, overlain by a shallow marine environment, and has since subsided more or less continuously to accommodate up to 12 km of sediment. The start of this subsidence is best seen in the Huqf Supergroup and end-Neoproterozoic salt basins [19, 20]. In contrast, shield regions were variably above sea level after ~750 Ma ago, and repeatedly uplifted and depressed, developing unconformities and continental sedimentary basins.

The eastern Arabian platform maintained marine to shoreline conditions for most of Phanerozoic time. Steady subsidence was periodically interrupted by block faulting and erosion (e.g., the uplift of the Central Arabian Arch and removal of Lower Paleozoic sediments [21], but subsidence always resumed. In contrast, the shield was possibly covered by fluviatile-shallow marine deposits during the Lower Paleozoic, but thereafter was largely emergent. From these observations, we infer that the differentiation of Arabian plate continental crust into uplifted shield and subsided platform began at ~750 Ma ago.

4. Geophysical Insights into the causes of Arabian Plate Uplift and Subsidence

4.1. Crustal Thickness & Velocity Structure

At issue here is the need to establish the typical thickness of Arabian plate continental crust and whether there are any significant regional differences in structure and thickness of crust and lithosphere. Thickness estimates are made from crustal refraction, teleseismic receiver function, and surface wave studies. A major crustal refraction experiment was undertaken in 1978, a SW-NE oriented, 1000-km long profile from near Riyadh on the Arabian platform to the Red Sea [22, 23, 24, 25]. This showed the crust to range from 43 km thick near Riyadh to 38 km near the Red Sea (figure 2), with a significant change in inferred upper crustal structure on either side of CAMA. The crust is composed of two layers, each about 20 km thick, with average P-wave velocities of about 6.3 and 7.0 km/sec, respectively [22, 23]. This velocity structure indicates that intermediate to felsic upper crust overlies mafic lower crust across a well-defined Conrad Discontinuity. Support for the interpretation of mafic lower crust beneath the Arabian Shield comes from mafic lower crustal xenoliths brought up in Neogene lavas [26, 27, 28]. Teleseismic receiver functions recorded at 23 stations in Saudi Arabia and Jordan analyzed by [29] indicate that the average crustal thickness beneath the Arabian shield is 39 km, consistent with results from refraction studies. A surface wave study of the entire shield indicated that the Moho lies 41-46 km deep [30].

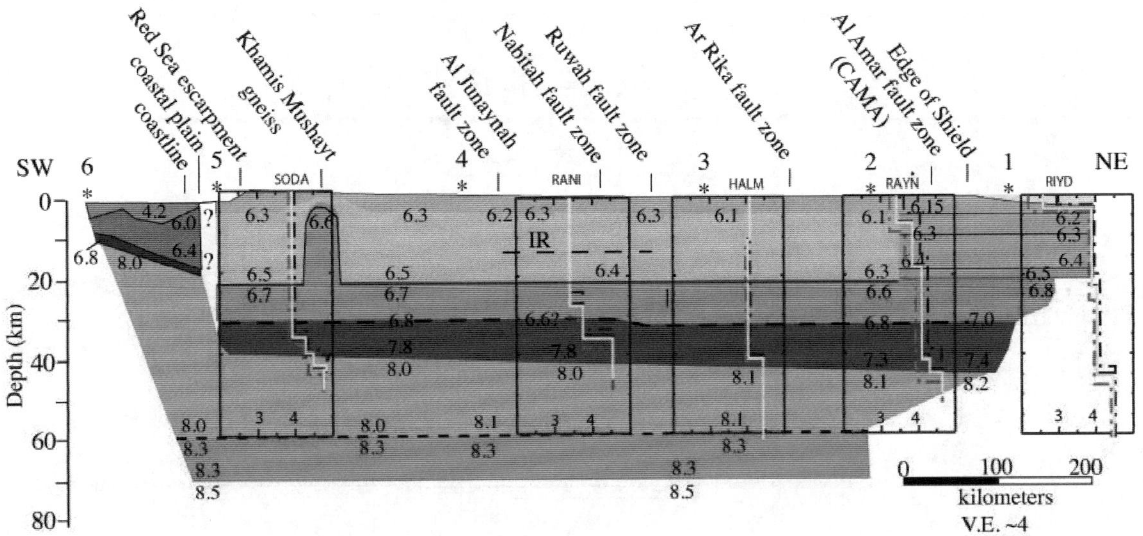

Figure 2. Interpretive crustal section for the 1978 seismic refraction profile across the Shield in southern Saudi Arabia, from the Red Sea to slightly east of the Arabian Shield [22]. P-wave velocities for each layer are given in km/sec. Numbers above asterisks = shot points. Overlain on this is shear-wave velocity structure from receiver function stacks of [31] for seismic stations at SODA, RANI, HALM, RAYN and RIYD. The yellow line is the optimal receiver function S-wave mode assuming a Poisson's ratio of 0.25; red and blues lines show 2 SD error ranges. Note similarities and differences between Moho determined by refraction to occur just above where Vp~8km/sec (boundary between brown to green) and receiver function inference drawn where Vs~4.6km/sec (short dashed line).

Crust beneath the northern Arabian plate is thinner, from 33 to 37 km, although a thickness of ~40 km was determined in the extreme north, just south of the Bitlis suture, according to receiver function studies [32]. Gravity models indicate that the crust beneath Syria is ~32 km thick [33]. combined gravity modeling and receiver function analysis to show that the crust of Syria north of the Palmyrides (figure 1) is significantly thinner (~25 km) than that to the south (~35km) [34]. Combined P-wave receiver functions and surface (Rayleigh) waves have been used to image the crust beneath two sites in Iraq: Mosul (near the Zagros Fold Belt) and Baghdad (in the Mesopotamian Foredeep). These results show that the crust is 39 km thick beneath Mosul and 43 km thick beneath Baghdad, and that crustal structure beneath these regions is similar [35]. These indications of variable (and often thinner) crustal thicknesses in the north may reflect the effects of Permo-Triassic and Mesozoic rifting associated with the formation of Tethys.

There are few geophysical studies of crustal and lithospheric structure beneath the Arabian Platform, but some evidence indicates different crustal compositions on either side of CAMA. [36] first showed evidence that the crust of the Arabian Platform was geophysically distinct from that of the Arabian Shield, using surface waves and teleseismic body wave arrivals. They found that the crust beneath the platform was slightly thicker (40km) and slower (Vp = 6.1 km/sec) than that beneath the shield and did not have a strong velocity gradient, consistent with an overall felsic crust for E. Arabia. In contrast, crust beneath the Arabian Shield is thinner (36km), ~6% faster, and shows a significant increase in velocity with depth, indicating mafic lower crust. It thus appears that the crust east of CAMA is slightly thicker and more felsic than that to the west, although more geophysical investigations are needed to test this idea. If confirmed, these results indicate that the crust east of CAMA should be elevated relative to the crust to the west, due to isostasy; this is opposite to what is observed, with the Shield elevated to the Platform.

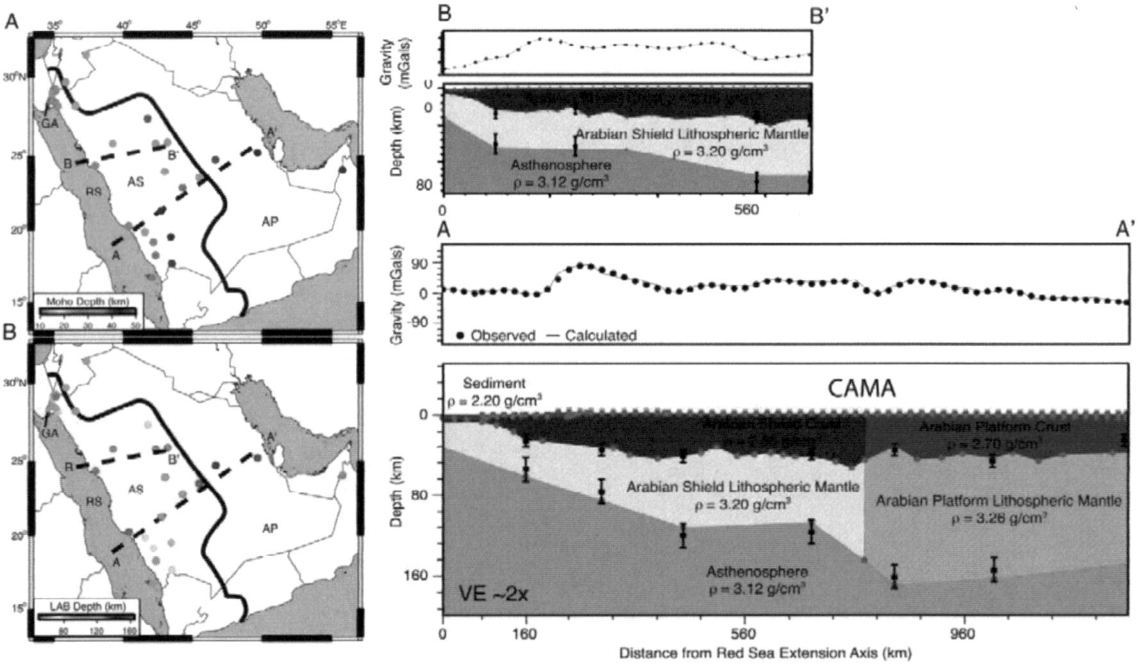

Figure 3. Thickness of crust and lithospheric mantle beneath western Arabia, modified after [37]. Colored circles on maps show the A) depth to Moho and B) depth to the lithosphere-asthenosphere boundary (LAB). The solid line approximates the boundary between the Arabian Shield (AS) and the Arabian Platform (AP) while the two dashed lines mark the locations of cross-sectional profiles AA′ and BB′. RS: Red Sea, GA: Gulf of Aqaba. Profiles on right show gravity signature and lithospheric structure along cross-sectional profile BB′ and AA'. Comparisons are shown between observed gravity from the GRACE satellites (black dots) and the calculated gravity (grey line) resulting from the lithospheric models shown below them. Red dots mark nodes that were used in the gravity modeling to constrain the boundary depths, and the densities (ρ) of each layer are listed. For stations along the profile, the Moho and LAB depths from receiver function analysis are shown by black squares with error bars. Errors on Moho and LAB depths are 5 and 10 km, respectively, and agree within error with the gravity model. Note that the jump in lithospheric thickness shown on profile A-A' occurs beneath CAMA.

4.2. Lithospheric Thickness

The thickness of Arabian plate lithosphere, or the depth to the lithosphere-asthenosphere boundary (LAB) is difficult to determine because velocity contrasts across this boundary are small. Hansen et al. [37] used S-wave receiver functions and supplemented these by modeling GRACE gravity to estimate the depth to the LAB (figure 3). They found that the LAB along profile AA' near the Red Sea coast lay ~50 km deep but increases to about 120 km beneath the Arabian Shield. A significant increase in lithospheric thickness is observed beneath the Shield–Platform boundary, where [37] show the LAB depth increasing to about 160 km. The results of Hansen et al (2007) showing thicker mantle lithosphere east of CAMA highlights its significance as marking an important lithospheric boundary.

5. Implications for the Evolution of the Arabian Continental Lithosphere and the Arabia-Eurasia Convergence Zone

Combined geologic and geophysical insights outlined above suggest five important points: (1) an important Neoproterozoic geosuture (CAMA) can be traced magnetically from central Arabia northwards; (2) that this separates crustal tracts with distinct Neoproterozoic histories (older in the east, younger in the west); (3) that E. Arabian (Platform) crust is slightly thicker and more felsic than that of W. Arabia (Shield) and thus more buoyant; (4) that E. Arabia is underlain by thicker lithosphere than that of W. Arabia; and (5) that W. Arabia has a long history of uplift whereas the platform has long subsided. Future studies are needed to confirm observations (2), (3), and (4), but if correct, these observations compel three conclusions. First, crustal variations cannot explain uplift of E. Arabia and subsidence of W. Arabia, because thicker, less dense E. Arabia crust should stand higher than thicker, denser W. Arabia crust due to isostasy. Second, regional variations in lithospheric thickness can explain uplifted Shield and subsided Platform as due to isostasy, because lithosphere (~3.3 g/cc) is denser than asthenosphere (~3.25g/cc), pulling down E. Arabia relative to W. Arabia. Third, regional variations in lithospheric thickness largely reflect the earlier stabilization of E. Arabia relative to W. Arabia. Continental Arabian plate mantle lithosphere began to form when igneous activity ceased, thickening with time due to conductive cooling, which began 150-200 Myr earlier in the east than in the west. This explains why Arabia east of CAMA was inundated by shallow seas ~540 Ma ago and accumulated salt whereas Arabia west of CAMA did not, as well as the long history of subsidence of the Arabian Platform and uplift of the Arabian Shield. These differences in lithospheric thickness may have been further modified by thermal erosion in Cenozoic time due to Red Sea rifting and the Afar hotspot.

We infer that E-W variations in Arabian Plate lithospheric thickness affected the Arabia-Eurasian convergence zone. Using the crustal and lithospheric thicknesses and densities of figure 3 (A-A'), Arabia west of CAMA has a density of 3.08 g/cc whereas to the east the density is 3.12 g/cc, making E. Arabia significantly easier to subduct than W. Arabia. In addition, the presence of Neoproterozoic salt in the east conceivably reduced friction on the plate interface, allowing Platform sediments to be scraped off to form the Zagros fold-and-thrust belt. In contrast, the thinner lithosphere of W. Arabia resulted in significantly lower densities in spite of the presence of denser crust in the west than in the east. The relatively buoyant lithosphere of W. Arabia is not easily subducted, leading to collision in the western part of the convergence zone.

6. References

[1] Bird P 2003 *Geochem. Geophys. Geosys.* **4** 1027 doi:10.1029/2001GC000252
[2] Allen M, Jackson J and Walker R 2004 *Tectonics* **23** TC2008 doi:10.1029/2003TC001530
[3] Vita-Finzi C 2001 *J Struct. Geol.* **23** 521–30
[4] Jacobs J and Thomas R J 2004 *Geology* **32** 721–24
[5] Nicholson P G, Janjou D, Fanning C M, Heaman L M and Grotzinger J P 2008 *GEO 2008: Middle East Geoscience Conference and Exhibition (Manama, Bahrain, 3-5 March 2008)*
[6] Miller N, Johnson P R and Stern R J 2008 *Arab J. Sci. Eng.* **33** 55–77
[7] Johnson P R and Stewart I C F 1995 *Tectonophysics* **245** 37–52
[8] Al-Husseini M 2000 *GeoArabia* **5** 527–42
[9] Doebrich J L, Al-Jehani A M, Siddiqui A A, Hayes T S, Wooden J L and Johnson P R 2007 *Precamb. Res.* **158** 17–50
[10] Johnson P R and Kattan F H 2008 *Arab J. Sci. Eng.* **33** 3–16
[11] Gass I G, Ries A C, Shackleton R M and Smewing J D 1990 Tectonics, geochronology and geochemistry of the Precambrian rocks of Oman (*Geology and Tectonics of the Oman Region*) eds A H F Robertson, M P Searle and A C Ries (London: Geological Society of London, Special Publication) **49** 585–99
[12] Mercolli I, Briner A P, Frei R, Schönberg R, Nägler T F, Kramers J and Peters T 2006 *Precamb. Res.* **145** 182–206
[13] Allen P A and J Leather 2006 *Precamb. Res.* **144** 167–198
[14] Pallister J S, Cole J C, Stoeser D B and Quick J E 1990 *Geology* **18** 35–39

[15] Roger J et al 1991 *Geologic Map of J'alan, sheet NF 40-8E, Scale 1:100,000, with Explanatory notes* Directorate General of Minerals, Oman Ministry of Petroleum and Minerals
[16] Würsten F et al 1991 The Uplift History of the Precambrian Crystalline Basement of the jabal J'alan (Sur, Area) (*Ophiolite Genesis and Evolution of the Oceanic Lithosphere*) eds T Peters, A Nicolas and R G Coleman (Dordrecht) 613–26
[17] Davies F B 1984 *J. Geology* **82** 37–53
[18] Stern R J 1994 *Annual Rev. Earth Planet. Sci.* **22** 319–51
[19] Husseini M I and Husseini S I 1990 Origin of the Infracambrian Salt Basins of the Middle East (*Classic Petroleum Provinces*) ed J Brooks (London: Geological Society of London Special Publication)
[20] Allen P A 2007 *Earth Science Rev.* **84** 139–85
[21] Alsharan A S and Nairn A E M 1997 *Sedimentary Basins and Petroleum Geology of the Middle East* (Amsterdam: Elsevier) 942 pp.
[22] Mooney W D, Gettings M E, Blank H R and Healy J H 1985 *Tectonophysics* **111** 173–246
[23] Gettings M E, Blank H R, Mooney W D and Healey J H 1986 *J. Geophys. Res.* **91** 6491–512
[24] Badri M 1991 *Tectonophysics* **185** 357–74
[25] Mechie J, Prodehl C and Koptschalitsch G 1986 *Tectonophysics* **131** 333–52
[26] McGuire A V and Stern R J 1993 *Contrib. Mineral. Petrol.* **114** 395–408
[27] Nasir S 1995 *Eur. J. Mineral.* **7** 217–30
[28] Al-Mishwat A T and Nasir S J 2004 *Lithos* **72** 45–72
[29] Al-Damegh K, Sandoval E and Brazangi M 2005 *Earth Planet. Sci. Lett.* **231** 177–96
[30] Mokhtar T, Ammon C J, Herrmann R B and Ghalib H A A 2001 *Pure Appl. Geophys.* **158** 1425–44
[31] Sandvol E, Seber D, Barazangi M, Vernon F, Mellors R and Al-Amri A 1998 *Geophys. Res. Lett.* **25** 2873–76
[32] Angus D A, Wilson D C, Sandvol E and Ni J F 2006 *Geophys. J. Int.* **166** 1335–46
[33] Al-Saad D, Sawaf T, Gebran A, Barazangi M, Best J A and Chaimov T A 1992 *Tectonophysics* **207** 345–358
[34] Brew G, Barazangi M, Al-Maleh A K and Sawaf T 2001 *GeoArabia* **6** 573–616
[35] Gök R, Mahdi H, Al-Shukri H and Rodgers A J 2008 *Geophys. J. Int.* **172** 1179–87
[36] Rodgers A J Walter W R, Mellors R J, Al-Amri A M S and Zhang Y-S 1999 *Geophys. J. Int.* **138** 871–78
[37] Hansen S E, Rodgers A J, Schwartz S Y and Al-Amri A M S 2007 *Earth Planet. Sci. Lett.* **259** 256–65

IOP Conf. Series: Earth and Environmental Science **2** (2008) 012006 doi:10.1088/1755-1307/2/1/012006

The accretionary model for orogenesis and its application to the evolution of the Aegean crust

Gordon Lister[1] **and Marnie Forster**
Research School of Earth Sciences, The Australian National University, Canberra 0200, Australia

E-mail: gordon.lister@anu.edu.au

Abstract. Both modern and ancient mountain belts have characteristic 'fingerprints' determined by the sequencing of the tectonic mode switches that took place during their evolution. Accretionary Tethyan style orogenesis is characterized by a sequence of "push-pull" tectonic mode switches associated with the accretion of a succession of continental ribbons. The origin of the extensional episodes can be found in the driving forces provided by rapid "roll-back" of adjacent subducting slabs. Such slabs appear to be created during (indentation-triggered) foundering of the marginal basins that typified ancient Tethys. We suggest individual accretion events were marked by short-lived episodes of high-pressure metamorphic mineral growth, followed by the development of km-scale extensional shear zones. The mode switches are often evident in tectonic sequence diagrams as F_R Δ SZ sequences, where F_R are recumbent folds, Δ a growth episode of metamorphic minerals, and SZ are extensional shear zones. Microstructures imply that mineral growth in the Δ events was static, or that they took place with such rapidity that deflections of fabric due to accumulating strain are not evident. Visually, the appearance of static growth was maintained. In our examination of the Cycladic Eclogite-Blueschist Unit at least three separate F_R Δ SZ sequences have been documented. The evolution of the Aegean crust was thus marked by a complexity that will not be unravelled without modern microstructurally focussed geochronology, and geospeedometry, in particular using the $^{40}Ar/^{39}Ar$ isotopic system.

1. Introduction

This talk was in part based on a manuscript that has since been submitted [1] and on [2]. In this short contribution we describe the derivation of some of the tectonic sequence diagrams that were discussed during the Harrington Symposium.

In [1] we argue that mountain belts have characteristic 'fingerprints' determined by the sequencing of the tectonic mode switches that took place during their evolution. In an orogen affected by tectonic mode switches, what happens is largely dependent on the magnitude and rate of lithospheric stretching or shortening before and after individual mode switches, and upon how rapidly motion was taking place, before and afterwards. These factors fundamentally influence the thermal evolution of an orogen, and thus directly or indirectly determine the processes that take place within it. For example, orogens affected by abrupt "push-pull" mode switches are characterized by high-pressure metamorphism whereas orogens affected by abrupt "pull-push" mode switches are

[1] To whom any correspondence should be addressed.

characterized by high temperature metamorphism, magmatism and anatexis.

This insight leads to different views as to the nature of orogenesis, and as to what distinguishes accretionary *versus* collisional orogens. For example consider the notion that orogens are constructed and that thereafter they collapse. This simple view can be refuted, in the process assigning terms such as "syn-orogenic" or "post-orogenic" to the dustbin. There can be several episodes during which an orogen is "constructed" just as there can be several episodes during which lithosphere-scale extension tears a newly created orogen apart, as is the case during Alpine orogenesis. Gravitational energy driven collapse of the orogen can occur continually throughout this process, and is not restricted to an artificially defined so-called post-orogenic phase.

Tethyan style orogenesis is characterized by a sequence of "push-pull" tectonic mode switches associated with the accretion of a succession of continental ribbons. The origin of the extensional episodes can be found in the driving forces provided by rapid "roll-back" of adjacent subducting slabs. Such slabs appear to be created during indentation triggered foundering of the marginal basins that typified ancient Tethys. We suggest individual accretion events were marked by short-lived episodes of high-pressure metamorphic mineral growth, followed by the development of km-scale extensional shear zones.

In the language of tectonic sequence diagrams [2] the mode switches are often evident in tectonic sequence diagrams as F_R Δ SZ sequences, where F_R are recumbent folds, Δ a mineral growth episode, and SZ are extensional shear zones. Microstructures imply that mineral growth in the Δ events was static, or that they took place with such rapidity that deflections of fabric due to accumulating strain are not evident. Visually, the appearance of static growth was maintained.

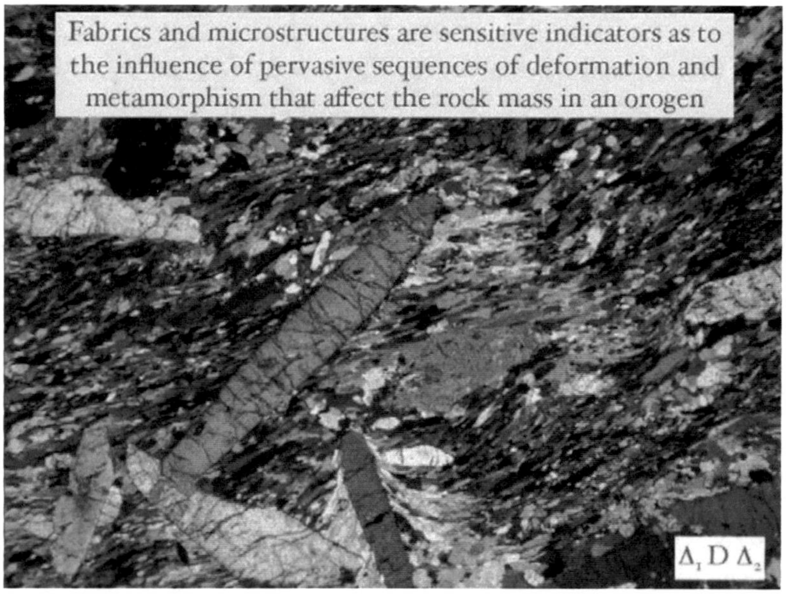

Figure 1. A thin-section of a blueschist from Vroulidia Bay, on the island of Sifnos, Cyclades, Greece. Blue amphibole has first grown (in Δ_1), and then been deformed. The deformation is denoted by D but in this case we could have equally well used SZ since we show this deformation occurred in a shear zone developed under blueschist facies conditions. This fabric has been statically overgrown by porphyroblasts of epidote (in Δ_2).

When we say that the growth of the epidote porphyroblasts in figure 1 took place under static conditions in fact we mean that no deflection of the inclusion trails within the porphyroblast can be observed under the microscope. Therefore the term "static growth" should not be taken to necessarily imply the absence of deviatoric stress during the growth of the porphyroblast. All that can be inferred from such microstructures is that growth was sufficiently rapid so as to take place without the accumulation of detectable strain. If growth was rapid (e.g., in a period $<<$ one million years) this condition can readily be fulfilled, even during moderately high strain rates. In contrast we might expect 3-4 million years to elapse between the growth of the blue amphibole and the overprinting porphyroblasts. It takes time for significant strain to accumulate, and the intensity of the glaucophane fabrics observed in this section require local stretching to exceed $\lambda = 3$.

In the same thin section there is abundant evidence that deformation continued after the period of metamorphic mineral growth. For example as shown in figure 2, the relict glaucophane fabric within the porphyroblasts is straight, whereas immediately adjacent to the porphyroblasts, the fabric has begun to develop a characteristic curvature, typical of the effect of ongoing stretching. It is also evident that collision of porphyroblasts has begun to take place, as a direct consequence of movement during ongoing deformation. Garnet porphyroblasts fracture and metamorphic reactions begin to take place driven by the ingress of fluid, which in turn is driven by the formation of microdilation sites.

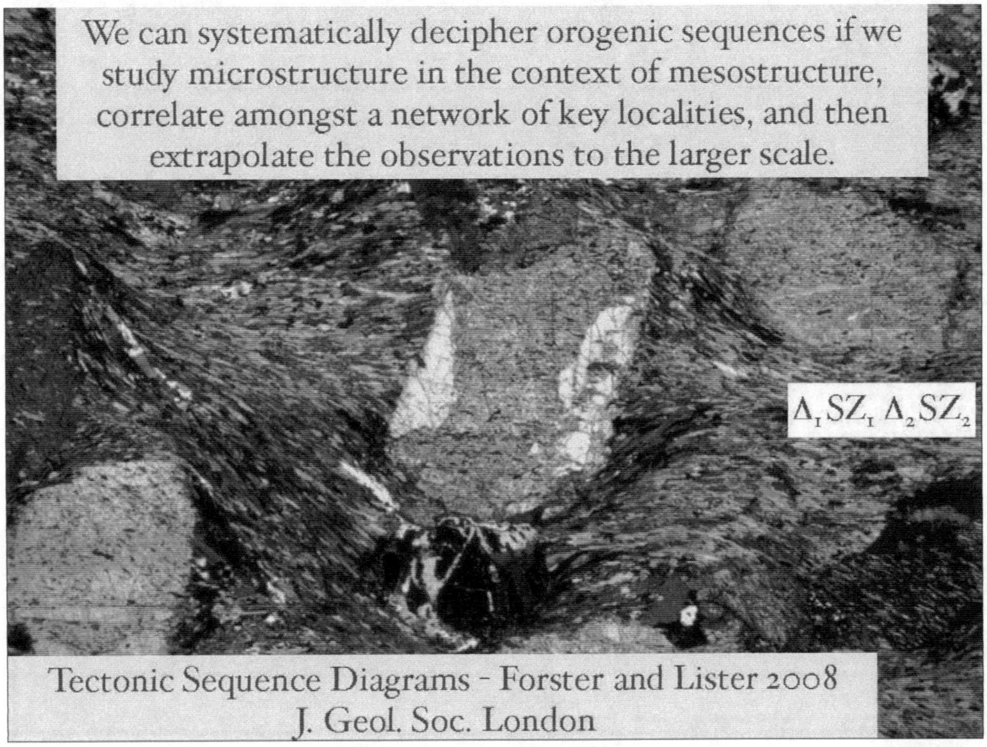

Figure 2. Microstructures in the same thin section from Vroulidia Bay, northern Sifnos, Further information as to the evolution of fabrics and microstructures can be discerned. Ongoing deformation after porphyroblast growth causes foliation to develop characteristic curvatures. A collision between a garnet (centre bottom) and an epidote porphyroblast (centre) in this time period triggers fracturing and ingress of fluid into microdilation sites. Pressure must have dropped, for the garnet begins to convert into albite.

This process leads to microscale shear zones, which can be seen in figure 2, emanating from the fracturing garnet clast, heading towards the top left of the microphotograph. Within microdilation sites related to the operation of this microshear it can be seen that chlorite, and then biotite, has begun to grow. Biotite rims chlorite, and chlorite rims garnet. Hence it can be concluded that the adjacent shear zone caused depressurization, and that thereafter these rocks were subject to a thermal pulse. The link between mineral growth an microdilation sites provides evidence for the involvement of fluids in this process. We must note at this point that we have evidence to suggest (in this case) that the same shear zone spasmodically operated through the time periods in which these microstructurally distinct and clearly distinguishable episodes of metamorphic mineral growth took place.

We can therefore add to the sequence diagram derived from that part of the thin section illustrated in figure 1, here interpreting the effects above as related to ongoing movement in a shear zone identified by mapping in the rocks adjacent to the sample in question. This data above means that our tectonic sequence diagram now becomes:

Δ_1 SZ Δ_2 SZ Δ_3 SZ
ep ab+gl Δ_4
chor $\rightarrow$ bio

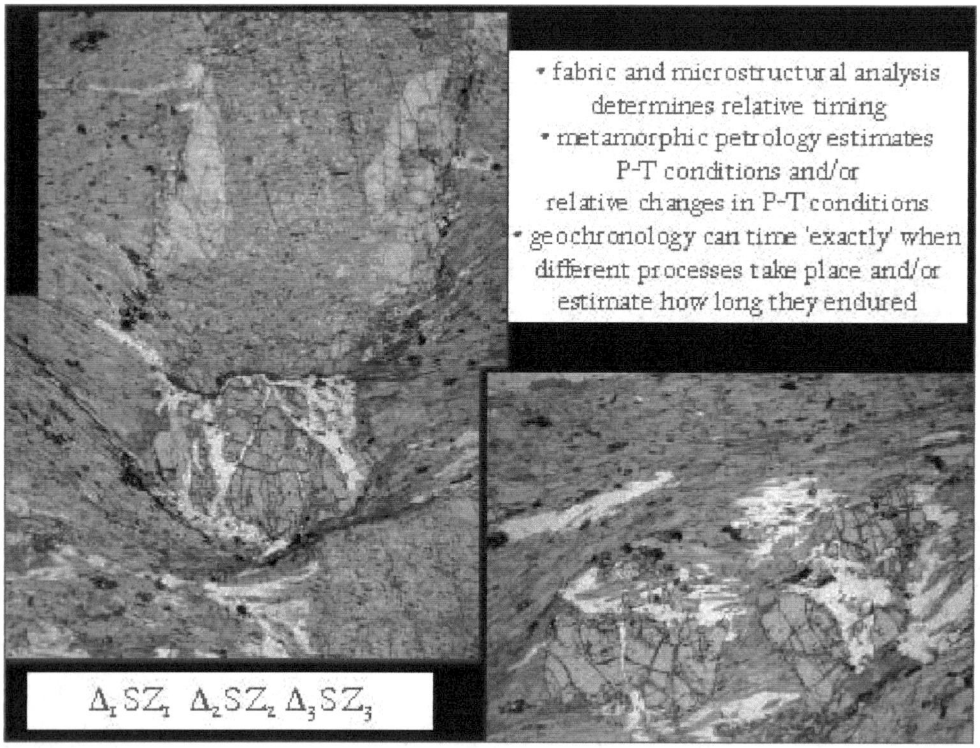

Figure 3. Detail of the microscopic processes accompanying the fracturing of garnet shows that quartz and glaucophane grew in the dilating pressure shadows adjacent to the garnet clasts, and this growth was synchronous with growth of quartz and albite. Overgrowth of (green) chlorite then took place, followed by biotite.

Closer examination of the mineral assemblages involved in each growth event allows identification of characteristic parageneses, as identified by [3]. The sequence diagram identified above can also be added to by fabric and microstructural analysis of folds in adjacent lithologies. The three periods of shear zone operation appear to be microstructurally distinct, and micro-structurally focussed $^{40}Ar/^{39}Ar$ geochronology [4, 5] allows the time periods involved to be ascertained. The sequence diagram thus becomes considerably more complex (figure 4).

It needs be emphasized that this is a work in progress. The conceptualization of the method of Tectonic Sequence Diagrams (TSDs) did not take place overnight, and although the method is now in routine use within our research group, in many aspects the observations that lead to the development of TSDs need to be performed anew, in the light of experience gained by their ongoing application.

$\Rightarrow D \Rightarrow\Rightarrow$ $\Rightarrow D \Rightarrow\Rightarrow$
Δ_{1A} S | Δ_{1B} S SZ | F_R Δ_{1C} SZ F_R Δ SZ
$>_{\Delta}>>>$ $>_{\Delta}>>>>>$ $>_{\Delta}>>>$
glaucophane omphacite glaucophane
lawsonsite jadeite garnet
D_1 $D_2 >>>$ $D_3 >>>>>>$

| F_R Δ_{1D} SZ | Δ_{2A} SZ F_R F_U
$>_{\Delta}>>>$
glaucophane albite
epidote biotite
D_4 D5

Figure 4. A tectonic sequence diagram compiled for Vroulidia Bay, based on observations such as made above (taken from [2] in press).

Our re-examination of the evolution of the Cycladic Eclogite-Blueschist Unit in this light has allowed the recognition of at least three separate F_R Δ SZ sequences, and these are now in process of being documented [6]. The evolution of the Aegean crust was thus marked by a complexity that will not be unravelled without modern microstructurally focussed geochronology, and geospeedometry, in particular using the $^{40}Ar/^{39}Ar$ isotopic system. Methods using differential geothermobarometry can be systematically applied to the evolution of particular microstructures, to determine whether shear zones are associated with pressure drops, for example, or whether the axial plane fabrics of recumbent folds are associated with pressure increases [cf 7]. While the methods of differential geothermobarometry do not allow absolute magnitudes of pressure and temperature to be estimated, they do allow important constraints to be added to the evolution of the pressure-temperature path, and thus to further qualify the evolution of the orogen in terms of a characteristic fingerprint, here characterized by F_R Δ SZ sequences.

2. References

[1] Lister G and Forster M 2008 Tectonic mode switches and the nature of orogenesis. *Lithos* in review
[2] Forster M A and Lister G S 2008 Tectonic sequence diagrams and the structural evolution of schist and gneisses in multiply deformed terranes. *J. Geol. Soc. London* in press
[3] Forster M A and Lister G S 2005 *Contrib. Mineral. Petrol.* **150** 523–45
[4] Forster M A and Lister G S 2003 *Aust. J. Earth Sci.* **50** 181–98
[5] Forster M A and Lister G S 2004 *J. Struct. Geol.* **26** 287–305
[6] Forster M A and Lister G S 2008 Core complex related extension of the Aegean lithosphere initiated at the Eocene-Oligocene transition. *J. Geophys. Res.* in review
[7] Beltrando M, Hermann J, Lister G S and Compagnoni R 2007 *Earth Planet. Sci. Lett.* **256** 372–88

IOP Conf. Series: Earth and Environmental Science **2** (2008) 012007 doi:10.1088/1755-1307/2/1/012007

Main geological problems of Western Anatolia and the significance of the Bodrum magmatic province

Yücel Yılmaz[1]
Professor, Kadir Has University, Cibali, Istanbul, Turkey

E-mail: yyilmaz@khas.edu.tr

Abstract. Western Anatolian Extended Terrain in Turkey stretches from the Balkan region in the north to the Taurides in the south. It contains a number of major tectonic entities including the Menderes Massif, the volcanic associations and the Neogene terrestrial cover sequence. In recent years the initiation of the N-S extension is viewed as a major factor responsible from the development of all these tectonic units. The initiation of the extension is regarded going back to the early Oligocene period. The data derived from our mapping project reveal that these units have developed under different tectonic regimes during different periods; for example the Menderes Massif, began to have formed during the late Creteceous and its development continued into the Miocene period. The Magmatic associations were formed in two separated phases; the early phase began during the Eocene- Oligocene time long before the extension started. The late phase is closely associated with the extensional regime. The Neogene sedimentary successions have 3 stratigraphic units separated by unconformities. The field data displays further that the N-S extension has not been uninterrupted. A major interruption occurred during the Early Pliocene period, and a region-wide, flat-lying erosional surface, as a key horizon, was developed.

1. Introduction

Western Anatolia is located at the eastern part of the Aegean extensional province. The following 4 major geological items of the region are still widely debated:

1. The Menderes Massif; its origin, mechanism of formation and age of development
2. The Magmatic Associations; their age and mechanism of formation
3. The Neogene Cover rocks; the tectonic regime under which they were developed and their tecto/ stratigraphic divisions
4. The N-S extensional regime; its time and mechanism of initiation.

A number of different views have been proposed on each one of these subjects. In this paper a review of the nature of these problems will be outlined together with some alternative solutions, in the light of data obtained from comprehensive mapping projects covering a considerably large part of western Anatolia from the Marmara Sea in the north down to the Mediterranean Sea in the south (figure 1).

2. The magmatic associations

[1] To whom any correspondence should be addressed.

The Magmatic Associations cover all of western Turkey from the Thrace Region and the Marmara region in the north down to the Bodrum Peninsula in the south (figure 1), and were formed in 2 discreet phases, the early phase and the late phase [1]. During the early phase, granitic stocks and small plutons, and intermediate and felsic volcanic rocks were extensively developed. The plutonic and volcanic rocks are closely associated in time and place (figure 1). Together they commonly formed in a collapsed caldera environment in which the centrally-located, deeper part is occupied by the granites, which are surrounded by the intimately-related volcanic suits, above the granites. These rocks are calc-alkaline in composition, and their compositions form a cluster displaying a common character and origin [1]. The composition of the volcanic rocks of this episode shift from calc-alkaline to shoshonitic trough time. The early phase of magmatic associations display magmatic arc geochemical signatures; the mantle-derived magmas were enriched by crustal components and later underwent AFC processes [1].

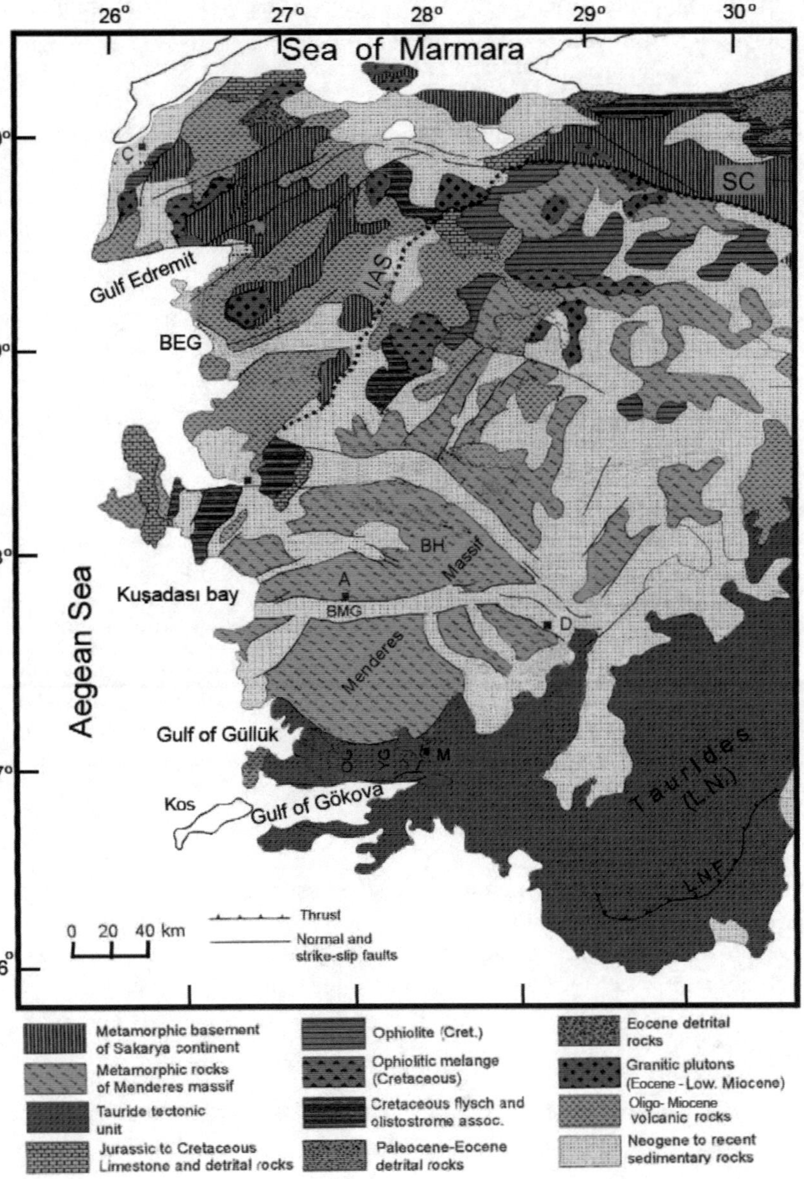

Figure 1. Geological Map of Western Anatolia, modified after [2, figure 1]. Abbreviations: IAS: Izmir-Ankara ophiolite suture; SC: Sakarya Continent; LN: Lycian Nappes; LNF: Lycian Nappe Front; BH: Bozdağ Horst; A, C, D, I and M are the cities of Aydin, Çanakkale, Denizli, Izmir, Muğla respectively; BEG: Bergama Graben; GDG: Gediz Graben; BMG: Büyük Menderes Graben; KT: Kale- Tavas basin; OG: Oren Graben; YG: Yatagan Graben.

The magmatic centers are aligned in NNE directions. The granitic magmas reached shallow levels in the crust along these trends and the volcanic rocks (figure 1), erupted along the faults and fissures defining the trend. A number of caldera type granites and associated volcanic centers have been identified in the region as exemplified from the Ezine and Kozak areas.

The late phase of the magmatic events produced mainly basic rocks which were missing during the early phase. They are sporadically developed and much less extensive. They form a distinctly different compositional cluster from the chemical composition of the early phase. Geochemically, the latter is alkaline in character, and displays similar affinities to the magmas that formed under extensional regimes [1]. The late phase of magmatic rocks began to form during the late Miocene, around 10 million years ago, and have continued till the present time. Therefore there is an apparent time gap between the development of the two magmatic phases. The gap is large in the northern regions where the initial phase began during the Eocene-Oligocene period. It narrows in the Bodrum area down to an interval of 1 to 2 million years.

The gradual migration of the volcanic front of the early phase displaying the arc magmatic affinities leads to the following assumptions:

1. They were formed in association with the subduction along the Hellenic trench. This subduction is assumed to have been continuing since at least the late Cretaceous period. This assumption is based on the length of the subducting slab which is more than 600 kms [3,4].
2. The southward migration of the arc magmas may be regarded as linked to the roll-back of the subducting slab; at least two stages of roll-back are observed along the length of the slab [4].

The late magmatic phase, which is closely related with the E-W trending faults and the associated structures, is considered to be the magmatic products which have formed under the extensional regime, when the N-S extension began [1].

This view relates the generation of the early phase and the late phase of the magmas to the different mechanisms, the subduction and the extension respectively, which explains the points listed below satisfactorily:

1. the migration of the volcanic front in time
2. the time gap between the two phases
3. the narrowing of the gap towards the south

3. The Menderes Massif

Under the name of the Menderes Massif is included a variety of metamorphic associations which crop out in the central and southern parts of western Anatolia occupying a 250 X 120 kms region (figure 1).

There is not an agreement among previous studies on most of the subjects related with the Menderes Massif, such as; its main components, their boundaries, their mechanism of development, etc. Generally the following zones or units (figure 2) are regarded as commonly associated with the Menderes Massif:

A. **The Afyon and Tavşanlı zones;** these two zones consist essentially of a metamorphosed sequence which are made up of three distinct layers [5]:
   A.1. A Paleozoic succession at the base, consisting of metaclastics,
   A.2. A thick carbonate sequence in the middle, which is Mesozoic in age, and is overlain tectonically by;
   A.3. an ophiolite nappe at the top.

These zones have suffered HP metamorphism during the latest Mesozoic-Early Tertiary period, under the thick (> 6 km) nappe pile.

B. **The Dilek Peninsula;** this zone is regarded as the easterly continuation of the Cyclades Massif of the Central Aegean regions, because they both have similar successions, and both have suffered similar metamorphic events including the HP metamorphism.

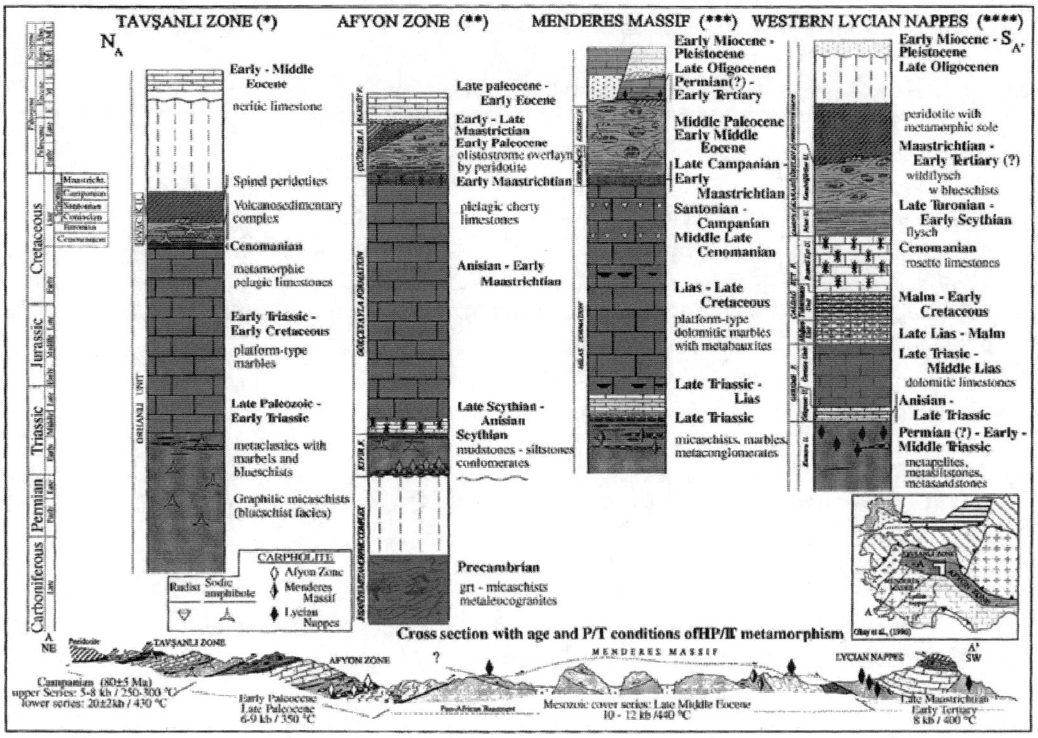

Figure 2. Generalized stratigraphic sections of the Menderes Massif and locations of HP/LT minerals. After [6, figure 8].

The main body of the Menderes Massif consists essentially of a Pan African basement association and its cover sequence [5] (figure 2). The former consists essentially of granites and gneisses, while the latter is made up of schists, phyllites and marbles. In places, the Pan African basement rocks are observed to have thrust above the cover rocks.

The southern zone is alternatively know as the Lycian Nappes [7, 8] (figures 1 and 2) and is commonly regarded as representing slices of the Taurus sequence (figure 2), which were transported southward as thrust sheets above the Menderes Massif during the development of the Orogeny [8] (figure 2).

All of the zones or units which form the Menderes Massif and its border zones may be regarded as belonging to the same entity, the Taurides (figure 2), because of their litho-statigraphic similarities and their ages [8, 9] (figure 2). They may be viewed as representing the northerly equivalent of the Tauride Carbonate platform facing the Tethyan Ocean, located in the north [8]. Following the total consumption of the ocean floor, the collision occurred between the Taurides and the Sakarya continent (figure 1). The underplated northern part of the Taurides underwent regional metamorphism, and formed the Menderes Massif when elevated [2, 6, 8, 10, 11]. Above the Menderes metamorphics, the southward transportation of the nappes took place from the late Cretaceous till the late Eocene, as evidenced by the deposition of the olistostromes, derived from the Nappes, which accumulated in front of the Nappe package [6, 8]. They were later overridden, and themselves underwent different degrees of metamorphism under the Nappes [2].

With respect to a centrally located horst (the Bozdağ horst) 3 geographic domains are usually distinguished in the Menderes Massif: the Nothern, Central, and Southern domains (figure 1). In the Central domain, the Bozdağ horst is delimited in the north and the south by low angle normal faults (figure 1). They are considered detachment faults. The horst is a young structure, elevated between these faults during the late Miocene period. The Central domain and the surrounding Northern and Southern domains share common lithological characteristics and metamorphic histories.

Three partly-coeval metamorphic events have been recorded in the Menderes Massif.

A. HP metamorphism ranging in age from 25 to 80 million years.
B. HT metamorphism ranging in age from 20 to 50 million years.
C. The cooling age of the metamorphic rocks, ranging from 5 to 50 million years. The three age groups get broadly younger toward the South. Collectively the age data indicate that:

C.1. Synorogenic events which affected the Menders Massif continued well into the Miocene period.
C.2. Exhumation of the Metamorphic associations occurred as a result of combinations of the following cooling paths [7]:
C.2.1. Cold path: rapid and steep path of ascent, along which HP metamorphic rocks were elevated [11]
C.2.2. Warm path; slow and low angle path of ascent, along which HP Metamorphic rocks were elevated.
C.2.3. The mixture of the two paths

Since the HP and HT metamorphic characters of these rocks survived during their ascents, the mechanism of elevation is assumed to have included a) high angle thrusting and back thrusting for the HP rocks, and b) low angle normal faults for the HT rocks.

4. The Neogene cover rocks

They are terrestrial deposits and form three tecto-stratigraphic units separated by unconformities [2]. The Lower Unit is early to middle Miocene in age, and was commonly deposited within NNE trending structural depressions lying sub-parallel to the volcanic axes. Away from the volcanic axes, this unit is represented commonly by fine-grained detritial rocks, formed in a low energy environment of deposition such as claystone, marl and fine-grained sandstone. Due to the proximity to the volcanic centers, the sediments alternate with, and include, increasing amounts of volcanoclastics, pyroclastics and lava layers [1, 2].

The Middle Unit is late Miocene-early Pliocene in age and is represented primarily by lacustrin white limestones. They are the most extensive Neogene rocks in western Anatolia, and apparently were formed in interconnected lake basins [2]. These shallow lakes appear to have covered all of western Anatolia during that period. Above the sequence of the Middle Unit, a region-wide flat-lying erosional surface can be observed, corresponding to a period of severe denudation which affected the region through the end of the early Pliocene. This surface may be used as a key stratigraphic horizon [2].

The Upper Miocene-Lower Pliocene limestone sequence and the erosional surface above that, is fragmented by E-W trending faults from after the early Pliocene period, when the present horst-graben system began to form. The infill of these grabens is fluvial sandstones and conglomerates [2].

5. The N-S extensional regime

In recent years the Aegean extensional region has come to be regarded as comparable to the Basin and Range region of the western United States [12, 13], and accordingly the Menderes Massif is evaluated as a core complex, formed under the extensional regime [14, 15].

The main geological characteristics of western Anatolia are dissimilar in nature to that of the Basin and Range. Some of the critical data for this may be listed as follows;

1. Changes in the modes of the magmatic activities forming two discreet phases correspond to the late Miocene. Initiation of the early phase began during the Eocene, long before the N-S extension began, at a time, when the region was still undergoing orogenic deformation. Therefore there is no one to one correlation between the initiation of the two events; the magmatism and the extension.
2. The Orogenic events which produced N-S compressional deformation continued interruptedly till the Middle Miocene. The Lower-Middle Miocene successions display many forms of shortening deformation such as folds, reverse faults, etc. Some pulses of release of the compressional regimes exemplified by the development of the region-wide, flat-lying erosional surface may correspond to the periods of roll-back of the subducting plate.

6. References

[1] Yılmaz Y 1989 An approach to the origin of young volcanic rocks of western Turkey *Tectonic Evolution of the Tethyan Region* ed A M C Şengör (Kluwer) 159–89
[2] Yılmaz Y, Genç C, Gürer F, Bozcu M, Yılmaz K, Karacık Z, Altınkaynak Ş and Elmas A 2000 *Geol. Soc. London Spec. Pub.* **173** 353–84
[3] Spakman W, Wortel M J R and Vlaar N J 1988 *Geophys. Res. Lett.* **15** 60–3
[4] Hinsbergen D J J , Hafkenscheid E, Spakman W, Meulenkamp J E and Wortel R 2005 *Geol. Soc. Am. Bull.* **33** 325–28
[5] Candan O, Dora O Ö, Oberhänsli R, Çetinkaplan M, Partzsch J H, Warkus FC, Dürr S 2001 *Int. J. Earth Sci.* **89** 793–811
[6] Candan O, Çetinkaplan M, Oberhänsli R, Rimmelé G and Akal C 2005 *Lithos* **84** 102–24
[7] Rimmelé G, Parra T, Goffé B, Oberhänsli R, Jolivet L and Candan O 2005 *J. Petrol.* **46** 641–69
[8] Şengör A M C and Yilmaz Y 1981 *Tectonophysics* **75** 181–241
[9] Şengör A M C, Satir M and Akkök R 1984 *Tectonics* **3** 693–707
[10] Ring U, Willner A P and Lackmann W 2001 *Am. J. Sci.* **301** 912–44
[11] Hinsbergen D J J, Zachariasse W J, Wortel M J R and Meulenkamp J E 2006 *Tectonics* **24**
[12] Wernicke B 1992 *Bull. Geol. Soc. Am* 553–81
[13] Dilek Y and Whitney D L 2000 Cenozoic crustal evolution in central Anatolia: extension, magmatism, and landscape development *Proceedings of the Third International Conference on the Geology of the Eastern Mediterranean: Geological Survey Department* eds I Panayides, C Xenophontos and J Malpas (Nicosia, Cyprus) 183–92
[14] Bozkurt E and Park R G 1994 *J. Geol. Soc. London* **151** 213–16
[15] Çemen I, Catlos E J, Göğüş D and Özerdem C 2006 *Geol. Soc. Am. Spec. Paper* **409** 353–79

IOP Conf. Series: Earth and Environmental Science **2** (2008) 012008 doi:10.1088/1755-1307/2/1/012008

Neogene volcanism and extension in Western Anatolian-Aegean area: A new geodynamic model

S Agostini[1,5], C Doglioni[2], F Innocenti[3], P Manetti[4] and S Tonarini[1]
[1] Istituto di Geoscienze e Georisorse, Consiglio Nazionale delle Ricerche, Via G. Moruzzi 1, 56124 Pisa, Italy
[2] Dipartimento di Scienze della Terra, Università La Sapienza, Roma, Italy
[3] Dipartimento di Scienze della Terra, Università di Pisa, Pisa, Italy
[4] Dipartimento di Scienze della Terra, Università di Firenze, Firenze, Italy

E-mail: s.agostini@igg.cnr.it

Abstract. The widespread Western Anatolian-Aegean Neogene volcanism presents a complex geochemical evolution reflecting the uncommon space-time variability of the geodynamic setting of the region. In the Western Anatolian and Central Aegean, a widespread supra-subduction magmatism, with calc-alkaline to shoshonitic affinity, took place from Early to Middle Miocene; this phase of activity ends with spots of ultra-K lavas and dykes. From Late Miocene onwards scattered alkali basaltic lavas with intraplate affinity were emitted, while calc-alkaline activity occurred in the South Aegean arc. Since Late Oligocene-Early Miocene, the region was, and still is, affected by extensional tectonics generally ascribed to a backarc rift. However the Aegean region should rather be considered as an unconventional backarc since its characteristics rather differ from "typical" backarcs. In fact, in spite of a long lasting(>40Ma) active NE-directed subduction of Africa, the backarc area still maintains a relatively thick continental crust (>20-25 km). Moreover, the upper Eurasian plate is overriding the lower Africa plate with separate segments, with Greece moving faster, and Turkey moving slower. The differential velocity between Greece and Turkey determines extension in the upper plate, unrelated to the loss of subducted retreating lithosphere, which is the usual setting for the origin of "classic" backarc settings. The geodynamic framework is supported by the geochemical and isotopic features of the supra-subduction magmas revealing the occurrence of a trapped, drying slab, with progressive decreasing of Fluid Mobile Elements/Fluid Immobile Elements ratios, $\delta^{11}B$ and $\delta^{7}Li$, coupled with scarce variations of Sr and Nd isotopes. Moreover, the differential motion between the Greek and Anatolian micro-plates creates tear zones with the formation of slab ruptures or vertical slab windows. The occurrence of such windows is, in fact, outlined by the presence of alkali basalts with intraplate affinity sourced in the sub-slab mantle.

1. Early Miocene-Pleistocene Volcanism in Western Anatolia

Complex magmatic activity took place in Western Anatolia in the last 30 million years (figure 1). The first volcanic cycle began with the emplacement of abundant lavas and pyroclastic products, which had a high-K calc-alkaline petrogenetic affinity and were mainly andesitic to rhyolitic in composition [e.g., 1, 2, 3 and references therein]. These products were shortly followed by shoshonitic volcanics, which are mostly Middle Miocene in age, and vary in composition from basaltic trachy-andesite to

[5] To whom any correspondence should be addressed.

rhyolite. The end of this magmatic cycle is marked by the emission of rare ultra-K products, both with Roman Magmatic Province-type or lamproitic affinity. Rocks belonging to the calc-alkaline, shoshonitic and ultra-K associations have the typical geochemical and isotope features of supra-subduction rocks: they are enriched in Fluid Mobile Elements (FME) with respect to High Field Strength Elements (HFSE), thus they have high Ba/Nb or Rb/Zr ratios. They also show high values of $^{87}Sr/^{86}Sr$ ratio (≈0.707-0.710), and low $^{143}Nd/^{144}Nd$ ratios (≈0.5122-0.5125) (figure 2).

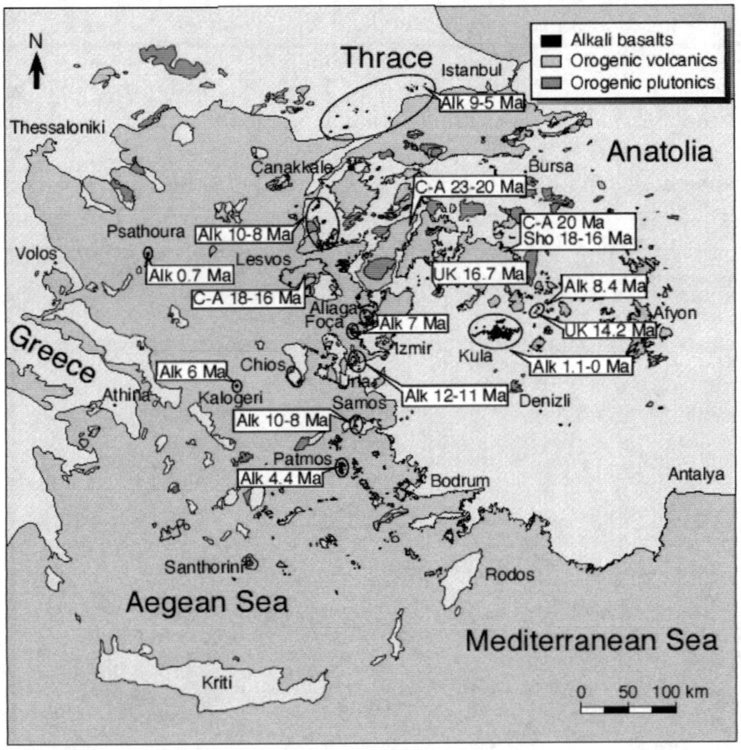

Figure 1. Sketch map of the study area, with magmatic rocks outcropping in the region. Age of volcanics are shown, alkali basalt outcrops are encircled. Redrawn after [4]. See [2 and 4] for age data sources.

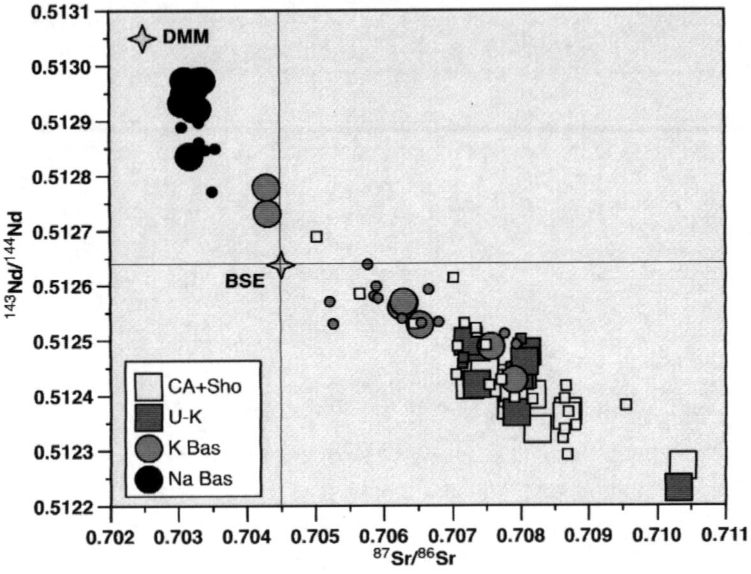

Figure 2. Sr and Nd isotope diagram of West Anatolia Neogene volcanic rocks (redrawn after [2]). CA+Sho, calc-alkaline and shoshonitic rocks; U-K, Ultra potassic lavas and dykes; K Bas, potassic alkali basalts; Na Bas, sodic alkaline basaltic rocks; DMM, Depleted MORB Mantle; BSE, Bulk Silicate Earth.

From Late Miocene to Pleistocene, scattered basaltic dykes and lavas were emitted [1, 2, 3, 4 and

IOP Conf. Series: Earth and Environmental Science **2** (2008) 012008 doi:10.1088/1755-1307/2/1/012008

references therein]. Some of these basalts are mildly SiO_2-undersaturated and have potassic affinity, whilst some others are strongly undersaturated and Na-alk. The K-basalts have very limited variability in major elements, and very high variability in trace element abundances, FME/HFSE ratios and Sr ($^{87}Sr/^{86}Sr \approx 0.704$-$0.708$) and Nd isotope values ($^{143}Nd/^{144}Nd \approx 0.5124$-$0.5129$). In contrast, greater major element variation is observed for the Na-alk rocks, which range to basanites to phono-tephrites, but have quite constant FME/HFSE ratios, $^{87}Sr/^{86}Sr$ (≈ 0.703) and $^{143}Nd/^{144}Nd$ (≈ 0.5129) (figure 2).

The source of calc-alkaline, shoshonitic and ultra-K magmas in Western Anatolia has been interpreted to be a mantle wedge strongly modified by a subduction component. The observed geochemical variability is considered to reflect mainly differing enrichments of subduction-related components, with a minor role played by the assimilation of crustal materials in the more evolved rocks [2]. The Na-alkaline basalts show the typical geochemical characteristics of intraplate magmas unaffected by subduction and they were sourced in the sub-slab asthenosphere.

These volcanic suites were the subject of a boron and lithium isotope study [5, 6], which provided fundamental insights on the mantle wedge modification processes. In the supra-subduction rocks a progressive lowering of $\delta^{11}B$ (and δ^7Li) is observed (figures 3a and 3b), matched by a progressive decrease of B/Nb ratio: this fact points to a decreasing amount of metasomatizing fluids in the mantle wedge, from a progressively dehydrating slab. In particular, the ultra-K rocks were sourced in a mantle wedge, left residual by the previous extractions of calc-alkaline magmas, then metasomatized by a very small amount of the latest fluids coming from an almost dehydrated slab.

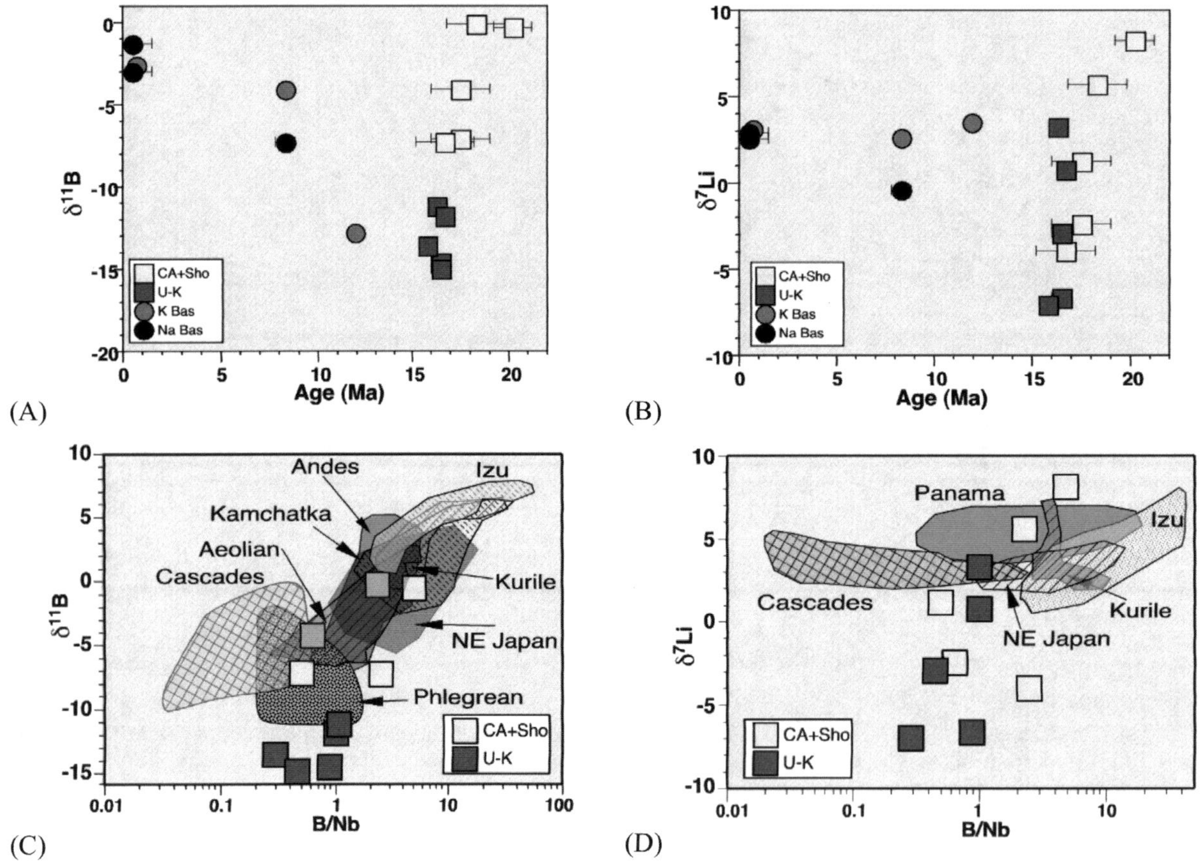

Figure 3. $\delta^{11}B$ and δ^7Li vs. B/Nb and age for West Anatolia Neogene volcanic rocks. Symbols as in figure 2. In (C) and (D), after [5], variation fields for selected worldwide arc lavas are reported.

The high $\delta^{11}B$ variability of K-alkaline basalts (figure 3a), along with their limited variation in major elements and wider variations of FME/HFSE ratios, Sr and Nd isotopes, point out that these basalts formed by interactions between sub-slab asthenospheric magmas and the latest slab fluids.

When comparing the B and Li isotope variation of West Anatolia supra-subduction rocks with worldwide arc lavas it is evident that the variation trend of B isotopes is rather common (figure 3c), even if most arc lavas exhibit a spatial (cross-arc) variation, whereas B isotopes variation in West Anatolia are observed not through space but trough time. Instead Li (figure 3d) isotope variability of this volcanic suite is unique.

We can assume that Li isotopes in downgoing slabs fractionate in the same fashion as B isotopes [7], and that a decoupling between B and Li occurs in the mantle wedge, because the mantle is bereft of B but contains significant Li, so that a Li isotope re-equilibration to common mantle values occur in the mantle wedge before arc magma genesis [8]. This happens in most arc lavas, but not in Western Anatolia, where a striking B-Li isotope co-variation is preserved.

2. What about extension in West Anatolia-Aegean region?

The Aegean region has long been described as a common backarc [e.g., 9], but it should rather be considered as an unconventional backarc since its characteristics and geodynamic constraints rather differ from "typical" backarc settings [10]. In fact, in spite of a long lasting (>40Ma) active NE-directed subduction of Africa, the backarc area still maintains a relatively thick continental crust (>20-25 km), unlike the fast "oceanization" of typical backarc basins associated to W-directed subduction zones.

Since [11], extension in Western Anatolia has been related to the Arabia-Eurasia collision but this model can be ruled out simply taking into account that extension in the region predates that collision and the velocity field (relative to Eurasia) is decreasing from west to east, that is moving toward the supposed energy source [12]. Furthermore, a westward extrusion of the Anatolia plate should rather generate compression in the Aegean area rather than extension.

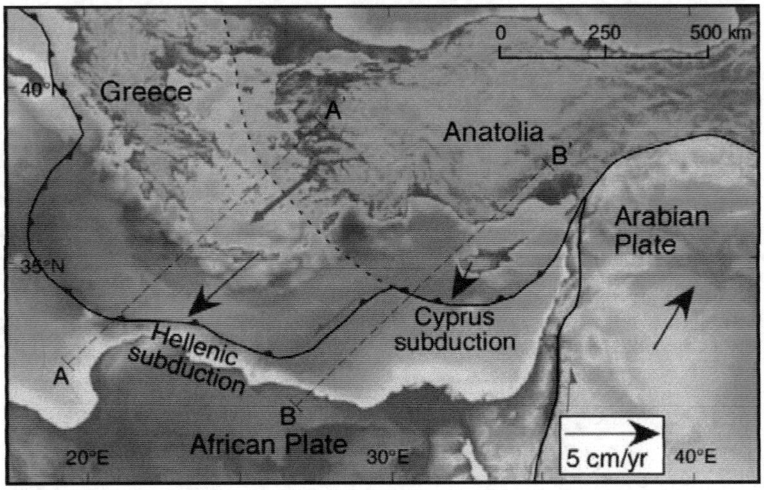

Figure 4. Tectonic framework of Eastern Mediterranean region (redrawn after [12]). Plate motions relative to fixed Africa are shown. Gray double arrow indicates the direction of extension. Margin between Greek and Anatolia microplates is indicative.

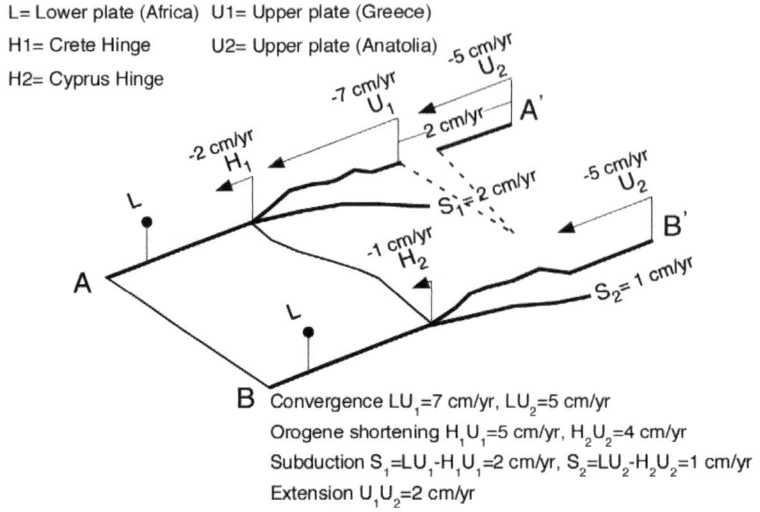

Figure 5. Schematic 3D diagram of Eastern Mediterranean region. Traces of AA' and BB' sections are reported in figure 4. Backarc extension occurs in Aegean area because of the different velocities of Greek (U1) and Anatolian (U2) microplates in overriding Africa.

Switching the reference frame, in the Eastern Mediterranean, the African plate is subducting under the Eurasian plate along the Crete and the Cyprus trenches. Considering fixed Africa, based on the GPS velocity fields [12] (figure 4) Greece and Anatolia act as two independent microplates: the differential velocity between Greece and Turkey in overriding Africa determines extension in the upper plate. This is a different geodynamic setting with respect to steep subduction zones, where the fast slab retreat relative to the upper plate is compensated by the asthenosphere, which is the usual setting for the origin of "classic" backarc settings associated to W-directed subduction zones [10]. It is evident (figure 5) that extension in the Aegean area takes place because three plates (African, Greek and Anatolian) are involved, and there is a different convergence speed between the two upper plates. No backarc extension is present in Cyprus-Central Anatolia where only two plates are involved in subduction.

3. Concluding remarks

3.1 Petrologic constraints to geodynamic evolution

Some important constraints on the geodynamic evolution of the area can be gained from the petrographical, geochemical and isotopic study of this volcanic suite:

(i) The Early-Middle Miocene evolution from calc-alkaline to ultra-K magmas implies that the genesis of orogenic magmas occurred by partial melting enduring on the same mantle domain,suggesting the occurrence of a non-convective mantle wedge under West Anatolia-Aegean.

(ii) The progressive decline of abundance of slab released fluids coupled to extreme B-Li negative signature implies that the slab was dehydrated up to almost dewatered.

(iii) The youngest Na-alkaline magmas were sourced in the sub-slab asthenosphere and suffered no interaction with slab fluids or any slab component. The genesis of such magmas is linked to extensional dynamics and asthenosphere upwelling; in addition, the occurrence of a slab window is invoked to allow such magmas to reach the surface.

(iv) The Late Miocene potassic-alkaline basalts do not derive from mixing between supra-and sub-slab magmas but they resulted from interaction between sub-slab magmas with residual slab fluids. This means that, during the earlier stages of slab window opening, limited interaction between intraplate magmas and slab fluids could occur.

3.2. Geodynamics and magmatism

A new geodynamic model (figure 6) was built merging together the observations from geodesy, geophysics, structural geology and the constraints coming from magma geochemistry.

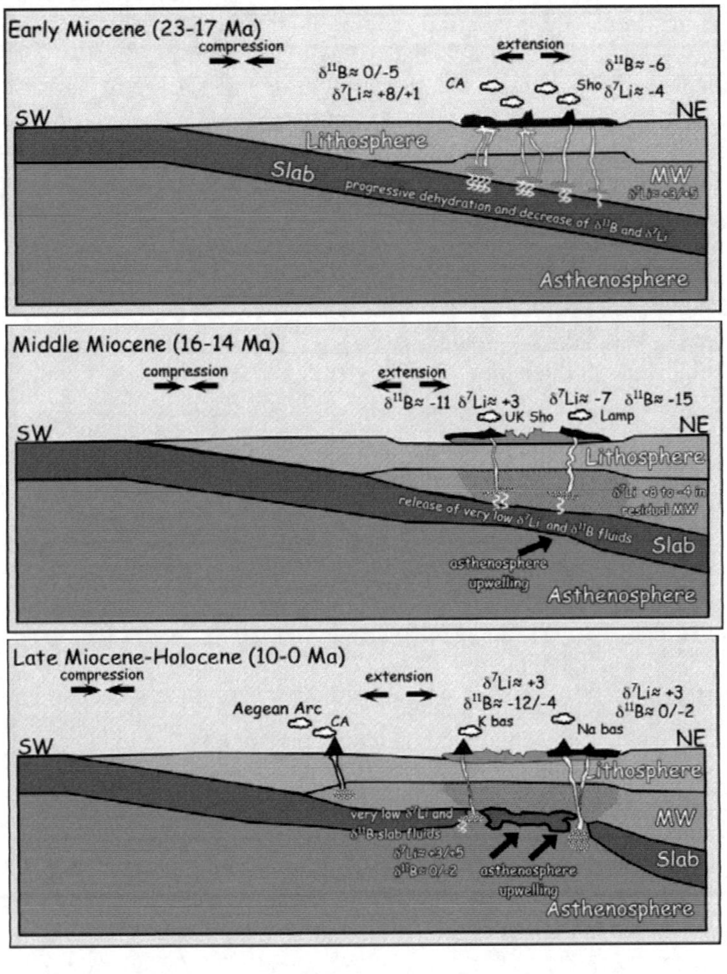

(A) (B) (C)

Figure 6. Schematic profiles (AA' profile of figure 4) of Western Anatolia and Aegean region illustrating the evolution of the geodynamic context, the processes leading to magma formation and Band Li isotope features of emplaced magmas.

(A) Early Miocene: slab-released fluids metasomatize the mantle wedge and activate the onset of widespread orogenic magmatism.

(B) Middle Miocene: the slab becomes to be stretched as a consequence of extensional tectonics and heated by the asthenosphere upwelling. Latest fluids released by a drying slab trigger partial melting in the overlaying mantle wedge, left residual by the previous magma extractions.

(C) From Late Miocene to Holocene, a new calc-alkaline activity settles on the South Aegean, whereas in the backarc the ongoing of extension gives rise to the formation of magmas with intraplate signature. These magmas are allowed to reach the surface after very scarce or no interaction with the overlaying slab, which is stretched and torn.

The Aegean-Western Anatolia region is characterized by a shallow and low dipping subduction zone [13]. The seismicity indicates an almost flat slab at 100-150 km depth in the central-northern Aegean Sea [13], with a severe and poorly comprehensible mismatch with the supposed slab inferred by mantle tomography, which is indicated as steeper and deeper, entering down into the lower mantle [e.g., 14]. Since hypocenter locations are more reliable than the model-driven mantle tomography, we suspect that the slab imaged by tomography is probably a misleading picture. Moreover, the African lithosphere subducting beneath Greece and Turkey is moving SW-ward relative to the mantle [15]. This indicates that the slab is moving toward the direction opposed to the subduction, consistent with the down dip extension focal mechanisms of the intra-slab seismicity [10]. Subduction is explained by the faster southwestward motion of the two upper plates, but the sinking of the Africa plate is very slow (figure 5) and measured by the convergence rate minus the velocity of the subduction hinge.

The subducted slab is progressively heated by the upwelling of the mantle beneath and dried by metamorphic reactions. Slab released fluids trigger calc-alkaline magmatism in the mantle wedge in Early Miocene (figure 6a) and ultra-K magmatism in the Middle Miocene (figure 6b). The mantle wedge trapped between the slab and the upper plate is thin and stagnant. In this context, the slab fluids have only limited interaction with host rocks before partial melting begins, and Li-isotopes cannot re-equilibrate to common mantle values, so keep their original signatures [6]. From the Late Miocene onwards, the development of extensional tectonics led to the stretching and tearing of both Aegean-Anatolian lithosphere, the mantle wedge and the subducted African slab, and induced asthenosphere upwelling (figure 6c): this caused the formation of intraplate-type magmas which reached the surface after suffering no interactions (Na alkaline basalts) or small interaction (K alkaline basalts) with latest slab fluids.

4. References

[1] Fytikas M, Innocenti F, Manetti P, Mazzuoli R, Peccerillo A and Villari L 1984 in Dixon T E and Robertson A H F (eds) *Geol. Soc. London Spec. Pub.* **17** 687–99

[2] Innocenti F, Agostini S, Di Vincenzo G, Doglioni C, Manetti P, Savaşçin M Y and Tonarini S 2005 *Mar. Geol.* **221** 397–421

[3] Aldanmaz E, Pearce J A, Thirlwall M F and Mitchell J G 2000 *J. Volc. Geoth. Res.* **102** 67–95

[4] Agostini S, Doglioni C, Innocenti F, Manetti P, Tonarini S and Savaşçin M Y 2007, in Beccaluva L, Bianchini G and Wilson M (Eds) *Geol. Soc. Am. Sp. Paper* **418** 1–15

[5] Tonarini S, Agostini S, Innocenti F and Manetti P 2005 *Terra Nova* **17** 259–64

[6] Agostini S, Ryan J G, Tonarini S and Innocenti F 2008 *Earth Planet. Sci. Lett.* doi:10.1016/j.epsl.2008.04.032

[7] Zack T, Tomascak P, Rudnick R L, Dalpe C and McDonough W F 2003 *Earth Planet. Sci. Lett.* **208** 279–90

[8] Tomascak P B, Widom E, Benton L D, Goldstein S L and Ryan J G 2002 *Earth Planet. Sci. Lett.* **196** 227–38

[9] Le Pichon X and Angelier J 1979 *Tectonophysics* **60** 1–42

[10] Doglioni C, Carminati E, Cuffaro M and Scrocca D 2007 *Earth Sci. Rev.* **83** 125–75

[11] McKenzie D. 1972 *Geophys. J. Royal Astr. Soc.* **30** 109–85

[12] Doglioni C, Agostini S, Crespi M, Innocenti F, Manetti P, Riguzzi F and Savaşçin M Y 2002 *J. Virt. Expl.* **7** 117–31

[13] Bourova E, Kassaras I, Pedersen A H, Yanovskaya T, Hatzfeld D and Kiratzi A 2005 *Geophys. J. Int.* **160** 1006–19

[14] Piromallo C and Morelli A 2003 *J. Geophys. Res.* **108** 2065

[15] Gripp A E and Gordon R G 2002 *Geophys. J. Int.* **150** 321–61

Cenozoic extensional tectonics of the Western Anatolia Extended Terrane, Turkey

Ibrahim Çemen[1,6], Elizabeth J Catlos[1,2], Oguz Gogus[3], Emre Diniz[4] and Mete Hancer[5]
[1] Boone Pickens School of Geology, Oklahoma State University, 105 Noble Research Center, Stillwater OK 74078, USA
[2] University of Texas at Austin, Jackson School of Geosciences, Geological Science Dept., 1 University Station C1100, Austin, TX 78712-0254, USA
[3] University of Toronto, Department of Geology, Earth Sciences Centre, 22 Russell St. Toronto, Ontario, M5S 3B1, Canada
[4] Occidental Oil and Gas Corporation, PO Box 22757, Houston TX 77227, USA
[5] Pamukkale Universitesi, Muhendislik Fakultesi, Jeoloji Muh. Bolmu, Denizli, 20070 Turkey

E-mail: icemen@okstate.edu

Abstract. The Western Anatolia Extended Terrane in Turkey is located on the eastern side of the Aegean Extended Terrane and contains one of the largest metamorphic core complexes in the world, the Menderes massif. It has experienced a series of continental collisions from the Late Cretaceous to the Eocene during the formation of the Izmir-Ankara-Erzincan suture zone. Based our field work and monazite ages, we suggest that the north-directed postcollisional Cenozoic extension in the region is the product of three consecutive stages, triggered by three different mechanisms. The first stage was initiated about 30 Ma ago, in the Oligocene by the Orogenic Collapse the thermally weakened continental crust along the north-dipping Southwest Anatolian shear zone. The shear zone was formed as an extensional simple-shear zone with listric geometry at depth and exhibits predominantly normal-slip along its southwestern end. But, it becomes a high-angle oblique-slip shear zone along its northeastern termination. Evidence for the presence of the shear zone includes (1) the dominant top to the north-northeast shear sense indicators throughout the Menderes massif, such as stretching lineations trending N10E to N30E; and (2) a series of Oligocene extensional basins located adjacent to the shear zone that contain only carbonate and ophiolitic rock fragments, but no high grade metamorphic rock fragments. During this stage, erosion and extensional unroofing brought high-grade metamorphic rocks of the Central Menderes massif to the surface by the early Miocene. The second stage of the extension was triggered by subduction roll-back and associated back-arc extension in the early Miocene and produced the north-dipping Alaşehir and the south-dipping Büyük Menderes detachments of the central Menderes massif and the north-dipping Simav detachment of the northern Menderes massif. The detachments control the Miocene sedimentation in the Alaşehir, Büyük Menderes, and Simav grabens, containing high-grade metamorphic rock fragments. The third stage of the extension was triggered by the lateral extrusion (tectonic escape) of the Anatolian plate when the North Anatolian fault was initiated at about 5 Ma. This extensional phase produced the high-angle faults in the Alaşehir,

[6] To whom any correspondence should be addressed.

Büyük Menderes and Simav grabens and the high-angle faults controlling the Küçük Menderes graben.

1. Scope of work

The Western Anatolia Extended Terrane in Turkey is located on the eastern side of the Aegean Extended Terrane and contains one of the largest metamorphic core complexes in the world, the Menderes massif (figure 1). It has experienced a series of continental collisions from the Late Cretaceous to the Eocene during the formation of the Izmir-Ankara-Erzincan suture zone. Based our field work and monazite ages, we suggest that the north-directed postcollisional Cenozoic extension in the region is the product of three consecutive stages, triggered by three different mechanisms.

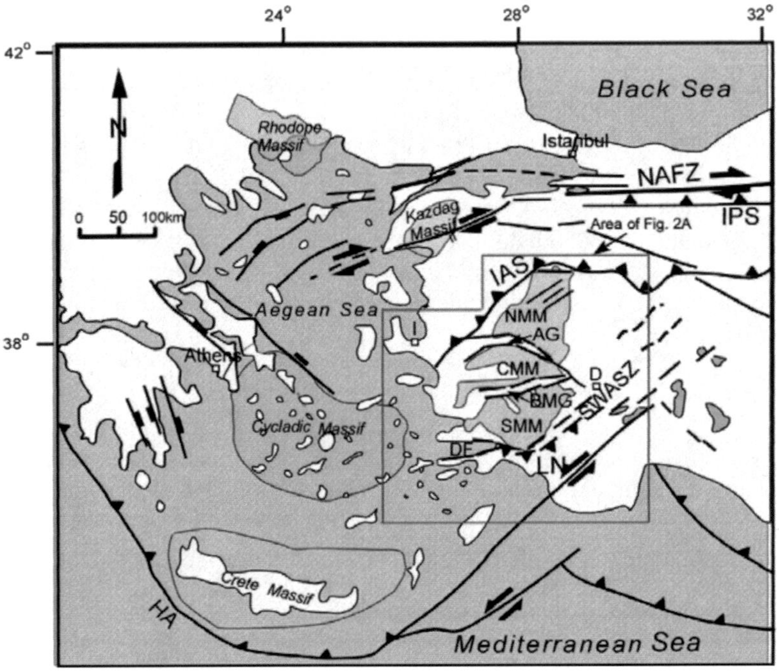

Figure 1. Generalized map of the Aegean Region showing the location of major structural elements and Alpine Metamorphic belts shaded in pink. The red box outlines the Menderes Massif in western Anatolia, Turkey (see figure 2). Abbreviations: Towns: D= Denizli, I= Izmir. Structural elements: AG= Alaşehir Graben; BMG= Büyük Menderes Graben; CMM= Central Menderes Massif; DF= Datca Fault; HA= Hellenic Arc; IAS= Izmir-Ankara Suture; IPS= Intra-Pontide Suture; LN= Lycian Nappes; NAFZ= North Anatolian Fault Zone, NMM= Northern Menderes Massif; SMM= Southern Menderes Massif; SWASZ= Southwest Anatolian Shear Zone.

The first stage was initiated about 30 Ma ago, in the Oligocene by the orogenic collapse of thermally weakened continental crust along the north-dipping Southwest Anatolian shear zone (figure 2a). The shear zone was formed as an extensional simple-shear zone with listric geometry at depth, and exhibits predominantly normal-slip along its southwestern end. But, it becomes a high-angle oblique-slip shear zone along its northeastern termination. Evidence for the presence of the shear zone includes (1) the dominant top to the north-northeast shear sense indicators throughout the Menderes massif, such as stretching lineations trending N10ºE to N30ºE; and (2) a series of Oligocene extensional basins located adjacent to the shear zone that contain only carbonate and ophiolitic rock fragments, but no high grade metamorphic rock fragments. During this stage, erosion and extensional unroofing brought high-grade metamorphic rocks of the Central Menderes massif to the surface by the early Miocene.

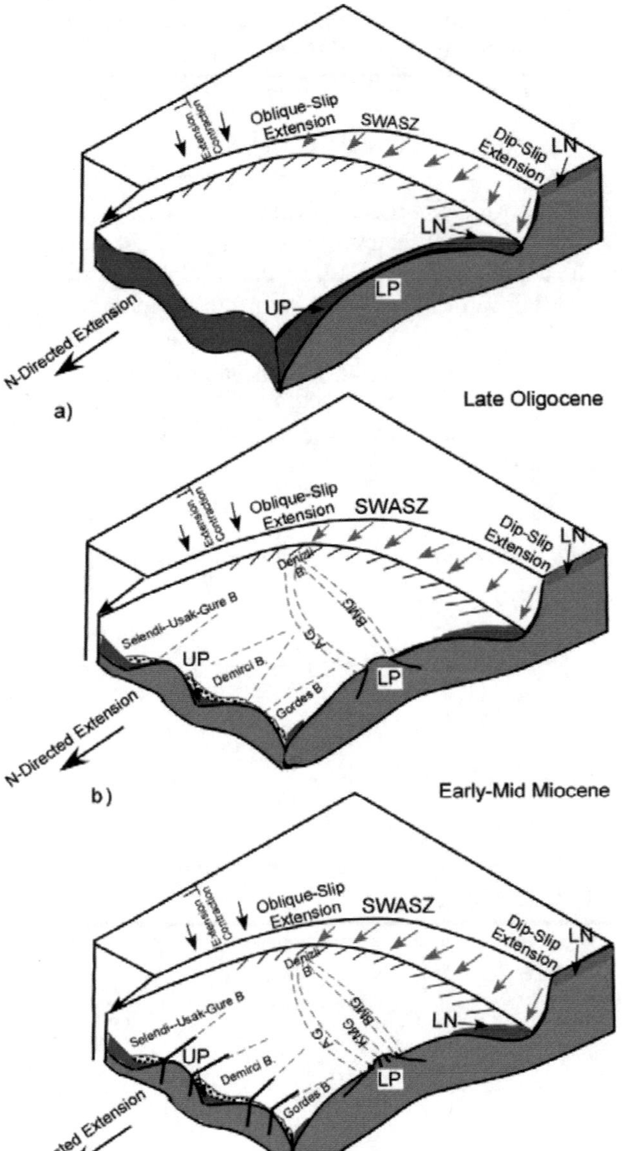

Figure 2. Three dimensional cartons showing structural evaluation of the Western Anatolia Extended Terrane during a) Late Oligocene; b) Early Miocene and c) Late Miocene to Pliocene. They are not drawn to scale and do not indicate the amount of extension in each stage. Abbreviations: AG= Alaşehir Graben; BMG= Büyük Menderes Graben; OB/KTB= Oren and Kale-Tavas Basins; LN= Lycian Nappes; KMG= Kucuk Menderes Graben; SG= Simav Graben; SWASZ= Southwest Anatolian Shear Zone [1].

The second stage of the extension was triggered by subduction roll-back and associated back-arc extension in the early Miocene and produced the north-dipping Alaşehir and the south-dipping Büyük Menderes detachments of the central Menderes massif and the north-dipping Simav detachment of the northern Menderes massif (figure 2b). The detachments control the Miocene sedimentation in the Alaşehir, Büyük Menderes, and Simav grabens, containing high-grade metamorphic rock fragments.

The third stage of the extension was triggered by the lateral extrusion (tectonic escape) of the

Anatolian plate when the North Anatolian fault was initiated at about 5 Ma. This extensional phase produced the high-angle faults in the Alaşehir, Büyük Menderes and Simav grabens and the high-angle faults controlling the Küçük Menderes graben (figure 2c)

The model outlined here can be tested by the following:

1) To carry out a quantitative study of the rocks in the Oligocene and Miocene basins of the WAET in terms of their tectonic and sedimentation history, including detailed observations on the lithologies of conglomerates. One critical required constraint is the timing of the appearance of high-grade metamorphic clasts in the conglomerates of the Oligocene to Miocene sedimentary successions in the Oren-Yatagan and Kale-Tavas Basins.
2) To conduct combined thermobarometric and geochronologic studies, such as in situ dating of monazite located in garnet-bearing assemblages used to generate P-T conditions. Currently, the peak metamorphic conditions reported for rocks of the Menderes Massif cannot be ascribed with confidence to specific tectonometamorphic events. In situ ion microprobe dating of monazite, and chemical correlation of those dates with silicate compositions and P-T conditions would provide insight into the nature of the complicated deformation events affecting Menderes Massif.
3) To conduct a petrologic and geochronologic study of the volcanic rocks well exposed throughout the WAET. If there is well-defined southward younging direction in the volcanic rocks of WAET, the timing and the role of subduction roll back in the Cenozoic extension in western Anatolia can be better understood.

2. References

[1] Çemen I, Catlos E J, Göğüş D and Özerdem C 2006 *Geol. Soc. Am. Spec. Paper* **409** 353–79

Tectonic block rotation, arc curvature, and back-arc rifting: Insights into these processes in the Mediterranean and the western Pacific

Laura M Wallace[1,3], **Susan Ellis**[1] **and Paul Mann**[2]
[1] GNS Science, P.O. Box 30368, Lower Hutt, New Zealand
[2] Institute for Geophysics, University of Texas at Austin, Austin, Texas, USA

E-mail: l.wallace@gns.cri.nz

Abstract. The fastest modern-day tectonic block rotations on Earth (up to 9 degrees/Myr) occur in the forearcs of convergent plate margins where a transition from collision of a bathymetric high to subduction of normal oceanic crust occurs. GPS techniques have enabled accurate documentation of the kinematics of these rotations, leading us to develop a conceptual model where the change from collision to subduction exerts a torque on microplates within the plate boundary zone, causing them to spin rapidly about an axis at the collision point. We have investigated geophysical and geological data from several active plate boundaries (from the western Pacific and Mediterranean regions) to document a compelling spatial and temporal relationship between the transition from collision to subduction, plate boundary curvature, and rapid tectonic block rotations. In some cases, these microplate rotations can initiate back-arc rifting. We also present numerical modelling results supporting our conceptual model for block rotations at collision/subduction transition. Our results suggest that the rate of microplate rotation depends on the incoming indentor velocity, and can be greatly enhanced by: (1) extensional stresses acting at the subduction interface (possibly due to slab roll back), and (2) a low-viscosity back-arc. Where viscosity of the back-arc is low, forearc microplate rotation dominates. In contrast, tectonic escape of strike-slip fault-bounded microplates is predicted in areas where the back-arc viscosity is high. Previous workers have suggested that the kinematics of the Anatolian block and back-arc rifting in the Aegean are influenced by some combination of forces associated with Arabia/Eurasia collision, and/or subduction (including slab rollback) at the Hellenic trench. Based on previous work from active western Pacific arcs, we propose that the collision of two separate indentors (Arabian promontory in the east, Apulian platform in the west), is a fundamental tectonic mechanism for large-scale anti-clockwise tectonic rotation of Anatolia and the opposing clockwise rotation of western Greece documented from paleomagnetic studies. The recognition of several global analogues for Mediterranean active tectonics may lead to new insights into the dominant forces behind tectonic processes there.

1. Subduction/collision-inducted microplate rotations in the Western Pacific

GPS velocities have recently documented rapid rotation of forearcs [1] in Papua New Guinea (figure 1), New Zealand (figure 2), Vanuatu, Tonga, and the Marianas. All of these regions are associated with marked plate boundary curvature and back-arc rifting, and are characterized by a transition from

[3] To whom any correspondence should be addressed.

collision of a buoyant indentor with the margin to nearby subduction oceanic crust of normal thickness. Modeling of GPS velocities [1] indicate that the forearcs in these locations spin as independent microplates (relative to the subducting plate) about the impact point at or near where the collision occurs, suggesting that the collisional indentor is a primary cause of the tectonic rotation. In some cases, back-arc basin formation occurs as a consequence of the microplate rotation [1] (figure 1).

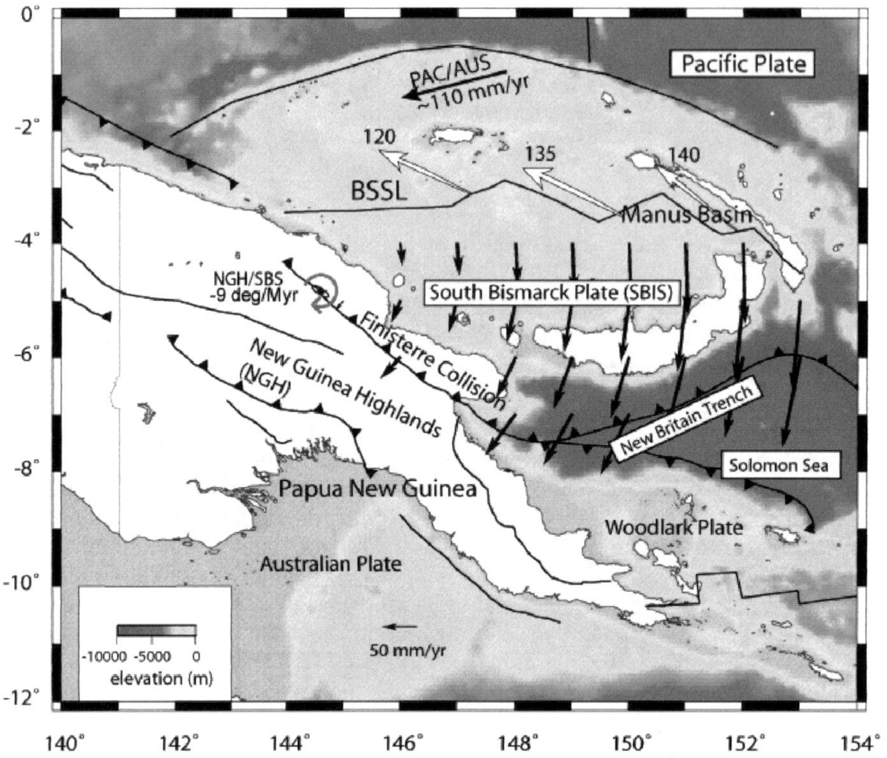

Figure 1. Tectonic setting of Papua New Guinea and vectors showing rotation of South Bismark plate (SBS) relative to the lower (New Guinea Highlands/Australia, NGH) plate (vectors are derived from GPS velocities, with elastic effect from interseismic fault coupling removed; [2]). The southern boundary of the SBS plate changes from collision between the Finisterre Range and underthrusting New Guinea highlands to northward subduction of oceanic crust in the Soloman Sea at the New Britain Trench. Pole of rotation for SBS microplate relative to NGH [2] shown by red semicircular arrow (black ellipse shows uncertainty in pole location), negative numbers indicate clockwise rotation (in °/Myr). Rates of back-arc rifting in the Manus Basin and strike-slip Bismark Sea seismic lineation (BSSL) are given in mm/yr (white arrows). The Finisterre collision began 3-3.7 Ma [3] simultaneously with the initiation of rifting in the Manus Basin at 3.5 Ma [4]. Paleomagnetic evidence indicates that the SBS plate has been rotating clockwise with an average rate of 8°/Myr since the Finisterre collision initiated [5], consistent with contemporary geodetic rates.

GPS velocities have recently documented rapid rotation of forearcs [1] in Papua New Guinea (figure 1), New Zealand (figure 2), Vanuatu, Tonga, and the Marianas. All of these regions are associated with marked plate boundary curvature and back-arc rifting, and are characterized by a transition from collision of a buoyant indentor with the margin to nearby subduction oceanic crust of normal thickness. Modeling of GPS velocities [1] indicate that the forearcs in these locations spin as independent microplates (relative to the subducting plate) about the impact point at or near where the collision occurs, suggesting that the collisional indentor is a primary cause of the tectonic rotation. In some cases, back-arc basin formation occurs as a consequence of the microplate rotation [1] (figure 1).

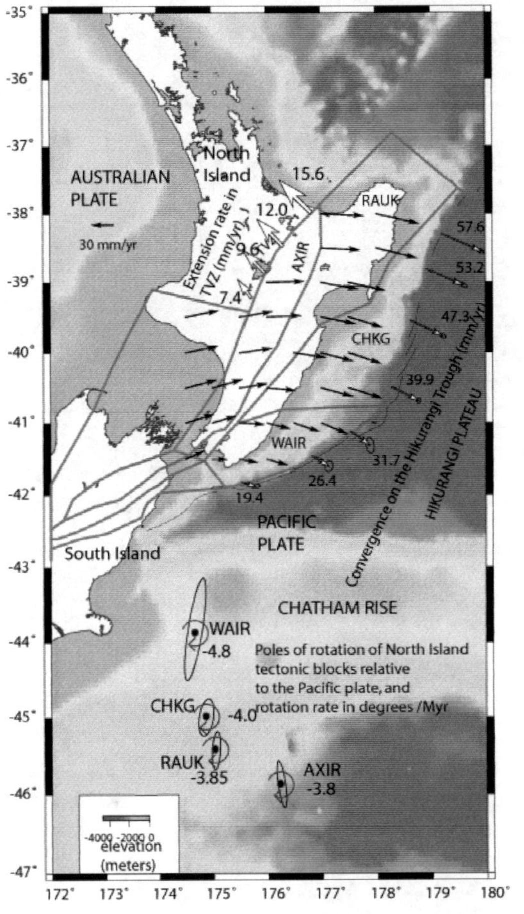

Figure 2. Tectonic setting of New Zealand showing rotating forearc blocks relative to the subducting Pacific Plate derived from modelling of GPS data and tectonic setting (see [6] for details of the modelling results). Rotating blacks and their respective poles of rotation (black ellipses show pole location uncertainties) are labelled (block names, WAIR= Wairarapa, RAUK= Raukumara, CHKG= Central Hikurangi, AXIR= Axial Ranges). White vectors show back-arc rifting in Taupo volcanic zone (TVZ) and black arrows show convergence rates at Hikurangi Trough (both labelled with rates in mm/yr). Arrows in TVS are not to scale. The Chatham Rise (a micro-continent on the subducting Pacific Plate) collides with the northern Southern Island; continental rocks on the subducting Pacific Plate transition northward to an oceanic plateau (the Hikurangi Plateau) which underthrusts the Northern Island at the Hikurangi Trough, while subduction of normal oceanic crust occurs at the Kermadec Trench further north. The eastern North Island pivots clockwise (at 3.8-5.0°/Myr) relative to the Pacific Plate about poles near the Chatham Rise collision point. Back arc rifting in the TVZ is a consequence of eastern North Island rotation [6], and most data suggest a wedge-shaped opening history for the TVZ [7, 8].

2. Conceptual model for block rotations at the collision/subduction transition

To explain the spatial correlation between rapid rotation of some forearc crustal blocks within plate boundary zones, and a nearby transition from collision to subduction at five locations in the western Pacific (figures 1 and 2), we [1] proposed a conceptual model where a torque exerted on the crustal blocks by the transition from collision to subduction causes rapid tectonic block rotation (figure 3). Where collision of a buoyant indentor occurs, the convergent plate boundary microblock is "pushed" into the upper plate, effectively transferring much of the plate boundary shortening into the upper plate. Away from the buoyant indentor collision, where subduction of normal oceanic crust occurs, the slab will either remain fixed in space (relative to the mantle) as it subducts (e.g., "slab anchor" model [9]), or it will roll back in the mantle, causing seaward retreat of the trench [10]. Trench suction forces (caused by subduction-induced flow patterns in the mantle wedge) move the upper plate toward the subduction zone. These suction forces also require the forearc to stay in contact with the subducting plate [11]. The landward "push" on the convergent plate boundary microplate where the buoyant indentor collision occurs, and the seaward "pull" where normal subduction occurs (due to trench suction and possible roll back) exerts a torque on the convergent plate boundary microblock. This torque causes the forearc to rotate rapidly about a pinning point relative to the subducting (lower) plate near where the buoyant indentor collision occurs, and slows convergence on the trench. If roll back of the slab occurs further along-strike (figure 3), this seaward "pull" force will be greatly increased, leading to faster forearc rotation rates. Our conceptual model does not explicitly require roll back of the slab, and therefore differs significantly from those models in which rollback is essential.

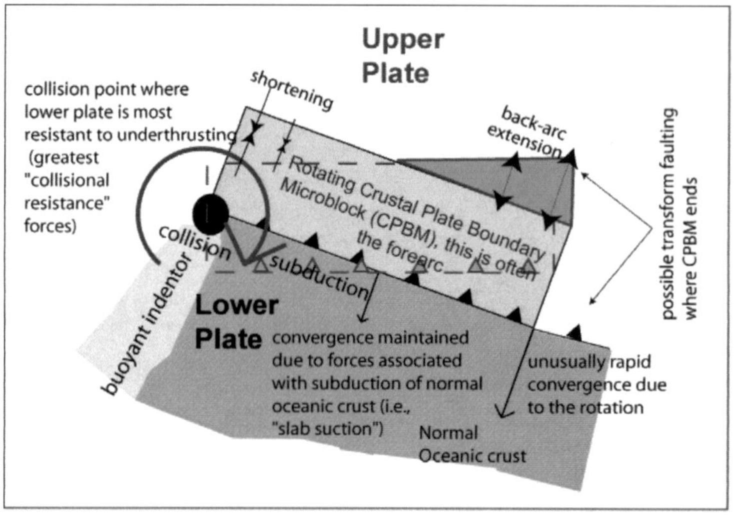

Figure 3. Detailed schematic of the collision/subduction-induced rotation model of [12] inspired by the western Pacific examples discussed in this paper.

A major prediction of the collision/subduction-induced rotation model is extension in the back-arc as the rotating crustal block pulls away from the upper plate. We suggest that rapid tectonic block rotations as a consequence of the lateral change from collision to subduction may lead to rapid but short-lived pulses of back-arc rifting. Moreover, our explanation that some back-arc rifts can occur as a consequence of collision/subduction-induced tectonic block rotation helps to reconcile the generally short-lived, episodic nature of most of the world's back-arc rifts, a common feature that two-dimensional, steady-state models for back-arc rifting fail to explain.

3. Numerical modeling of collision/subduction induced microplate rotation

We have illustrated the basic principle of collision/subduction-induced tectonic block rotation using a simple block-model representation of a subducting system [12] (figure 4). The model is deliberately kept simple, in order to identify its main physical mechanisms. Only the upper plate of the subduction system is modeled, as a Von Mises elasto-plastic body with a linear elastic strength. We impose a weak 100 km-wide back-arc zone, with a weakly frictional vertical interface to rotation and translation of the forearc microplate relative to the upper plate. The forearc microplate is also free to move at the top and bottom boundaries, (i.e., so the forearc microplate can detach from the upper plate). We simulate the effect of a buoyant indentor entering the subduction zone by imposing a compressional velocity of 1 cm/yr on the bottom right-hand edge of the model. Here, we show three different models: (1) buoyant indentor collision where stresses on the subduction boundary away from the collision point are neutral (i.e., no roll back occurs) (figure 4a), (2) buoyant indentor collision where extensional stresses are imposed on the subduction boundary (i.e., simulates the effect of slab roll back) (figure 4b), and (3) buoyant indentor collision where the viscosity in the weak back-arc zone is increased to 10^{24} Pas (figure 4c).

In the first model, rotation of the forearc occurs, suggesting that the torque induced by a simple collision/subduction transition (without slab rollback) is sufficient to produce tectonic rotation of the forearc block that may also lead to back-arc rifting (figure 4a). When viewing the model results relative to the lower plate, the forearc pivots about an apparent pole of rotation at the buoyant indentor collision point (at a rate of about 1 deg/Myr), where the velocity of the indentor is zero with respect to the forearc microplate, similar to the western Pacific examples discussed previously. The relative velocity between the forearc and the lower plate is increased at the northern end of the rotating forearc to 1.6 times of the rate of overall plate convergence (figure 1a), suggesting that the forearc rotation driven purely by a change from collision to subduction can produce a >50% increase in relative velocity. The rotation also leads to an along-strike change from shortening adjacent to the buoyant indentor, to extension in the back-arc further along strike (back-arc rifting at rates up to 6.4 mm/yr, or 0.64 of the buoyant indentor collision rate of 10 mm/yr). The rate of buoyant indentor rotation depends on both the length of the forearc microplate (shorter forearcs have higher rotation rates) and the incoming indentor velocity (higher indentor velocities lead to higher rotation rates) [12].

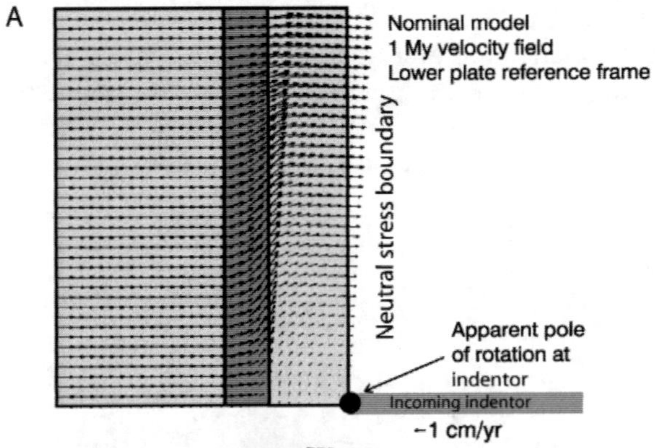

Figure 4: Resulting velocity field (after 1 Myr) with an indentation velocity of 1 cm/yr at the lower right corner, and (a) no extensional stress applied along the rest of the right hand (subduction) boundary, and (b) roll back, via an extensional stress of 1 MPa applied at the right-hand boundary, and (c) high viscosity in the backarc, which causes lateral escape rather than rotation. Note that the velocity fields in (b) and (c) are shown in an upper plate reference frame, while (a) is shown in a lower plate reference frame to demonstrate the pivoting of the forearc about the collision point. The forearc in (b) also rotates about the collision point when viewed relative to the lower plate.

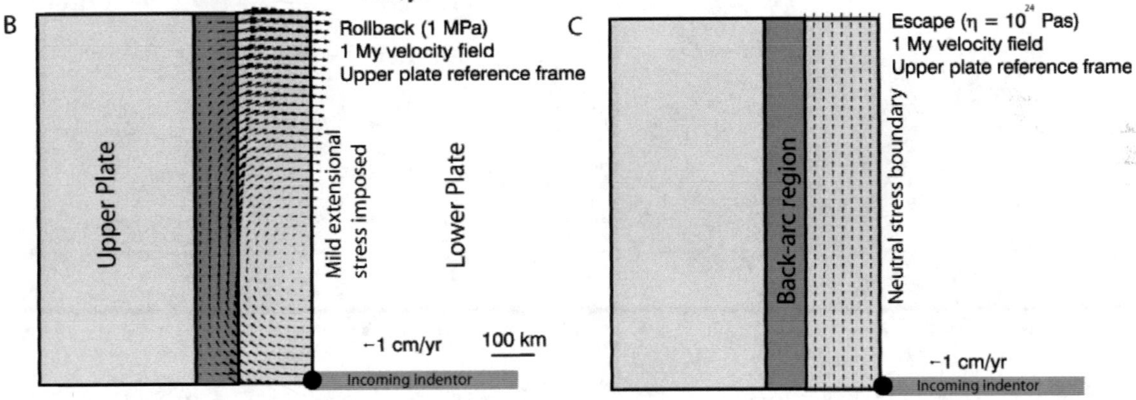

In the second model (simulating the influence of slab rollback), a very minor negative stress of 1 MPa is applied along the right-hand subduction boundary (figure 4b). In this case, the rotation rate increases by 3.7 times (compared to the non-roll back case), the total relative velocity between forearc and lower plate increases to 23 mm/yr (or, 3.1 times the upper plate/lower plate convergence rate), and the maximum rate of back-arc rifting goes up to 23 mm/yr (3.7 times greater than the non-roll back case). The outgoing velocity of the rotating forearc (relative to the upper plate) is greater than the incoming velocity of the buoyant indentor. For greater negative stresses along the subduction interface, even higher outgoing velocities can be attained.

The third model is identical to model one (e.g., a stress-neutral subduction boundary), but in this case we increase the viscosity of the back-arc region. This model shows that when the back-arc is too strong, the forearc is unable to rotate. In this case, weak strike-slip faults may dominate the system, leading to lateral tectonic escape. It is also important to note that the pure "tectonic escape" model does not produce any extension in the back-arc region.

4. Application of our conceptual model to the Mediterranean region

Similar to the western Pacific examples, the markedly curved arc systems and collided mountain belts in the Mediterranean (figure 5) record complex processes of back-arc rifting and ongoing, tectonic block rotations about poles of rotation sited near indentor collision points [13, 14, 15, 16, 17, 18]. Collision of the Arabian plate with Eurasia in eastern Turkey abruptly transitions to the west into north-directed subduction of the oceanic leading edge of the African plate at the Hellenic Trench, which in turn transitions northward into continent-continent collision of the Apulian platform with Albania. The pole of rotation of the anti-clockwise rotating Anatolian block relative to the Arabian plate--derived from GPS velocities[13,17]--occurs within 500 km of the Arabian collision point, similar to the western Pacific examples [1] (figure 5). Anti-clockwise rotation of Anatolia (1.2°/Myr from GPS, relative to Eurasia, 0.8°/Myr relative to Arabia [13]) is consistent with paleomagnetic studies [14] that indicate 25° of anticlockwise rotation (relative to Eurasia) since the Miocene (average rates of 2.5-1.25°/Myr if rotation started ~10-20 Ma). Although the exact timing of various tectonic events in the Mediterranean region is still under debate, there is a general coincidence between a middle Miocene Arabia/Eurasia collision[19, 20, 21, 22], and the tectonic rotation of Anatolia (the onset of which may be marked by the mid-late Miocene start of activity on the North Anatolian Fault [22, 23]).

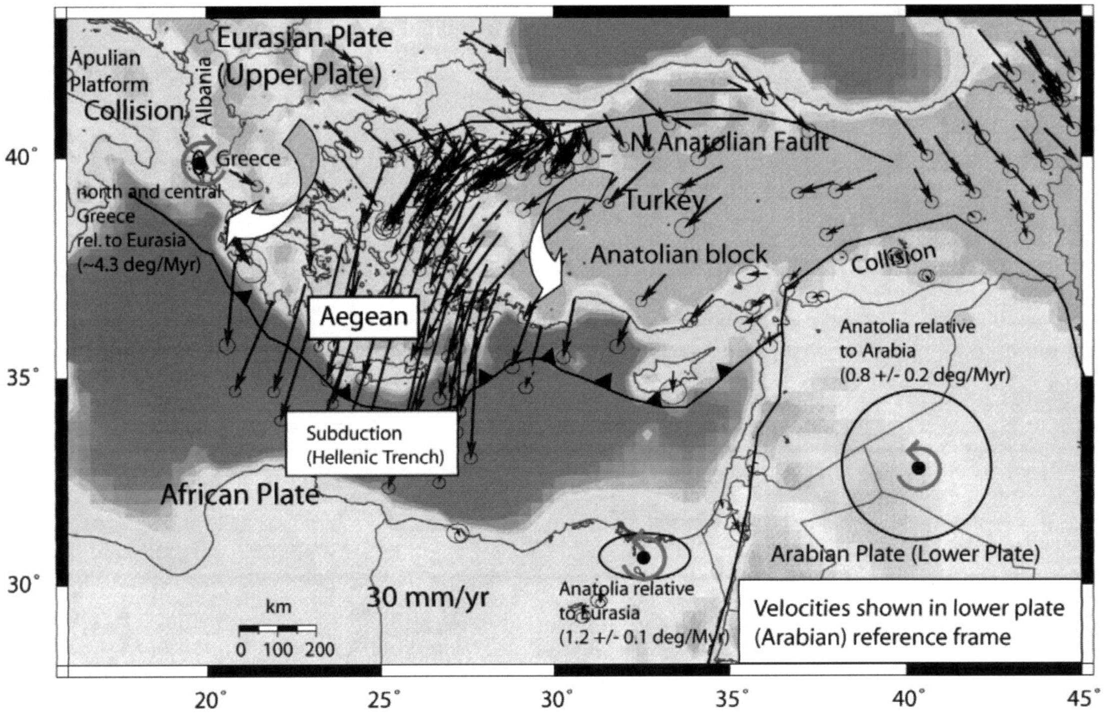

Figure 5. Tectonic setting of the Mediterranean region and GPS site velocities showing the kinematics of Turkey and Greece relative to the lower (Arabian) plate [13]. Poles of rotation (red, semi-circular arrows with black error elipses) for Anatolia relative to Eurasia and Arabia from [13], and for northern and central Greece relative to Eurasia [15, 17] indicated on map. Large white circular arrows schematically demonstrate the opposing rotation of Anatolia and western Greece.

In the western Mediterranean, the Apulian platform collides with Albania, while oceanic crust underlying the Mediterranean subducts beneath western Greece further south. Paleomagnetic data suggest that western mainland Greece and much of the Peloponnesian Peninsula have undergone ~40 degrees of clockwise rotation since the Miocene, about a pole of rotation near the Apulian Platform collision point [14, 16]. Currently, GPS velocities indicate that northern and central Greece pivot clockwise relative to EUR at ~4.3 °/Myr about a pole of rotation near the Apulian platform collision [15, 17] (figure 5). Rapid clockwise rotation of western and northern Greece appears to be temporally related to the Apulian platform collision, although there is some uncertainty with regards to the precise timing of its collision and rotation. Van Hinsbergen et al. [16] prefer a Late Miocene age for onset of Apulian collision; this would be just prior to Kissel et al.'s [14] 5 Ma start for clockwise rotation, but would postdate the start of major rotations ~15-8 Ma as inferred by van Hinsbergen et al. [16].

Overall, we suggest that the opposing clockwise and anti-clockwise tectonic rotations in the Mediterranean are dictated by the collision of two buoyant indentors with the Hellenic subduction zone: the Arabian plate and the Apulian platform. Conceptually, the kinematics of our "dual indentor" model is similar to the pinned, broken slat model [24], and the "double-saloon-door" model [18] which produces the "gigantic kink" of the Mediterranean margin.

The collision/subduction-induced block rotation model cannot explain all aspects of Mediterranean tectonics. For example, the trenchward movement of much of the Aegean region (with little associated modern-day rotation) requires a strong slab rollback influence [17]. The apparent lateral "escape" of Anatolia suggest that the kinematics of the Anatolian block may be influenced to some degree by tectonic escape processes as suggested by many previous workers (in addition to collision/subduction-induced microplate rotation that we suggest here). The assertion that the Black Sea region (just north of the north Anatolia Fault) is very strong and acts as a rigid back-stop [13] is consistent with our numerical models of tectonic escape where there is high back-arc viscosity (figure 4c). It is likely that a combination of processes (collision/subduction induced microplate rotation, slab rollback, and lateral tectonic escape) act together to influence the geodynamic framework and kinematics of deformation of the Mediterranean region.

5. Discussion/conclusions

Striking similarities in patterns of tectonic rotations and deformation in the western Pacific and Mediterranean region suggest that many of these areas are evolving by a common tectonic process that differs from place to place only by its spatial scale and evolutionary stage of development. We have shown that rapid tectonic block rotations (and commonly associated back-arc rifting) are associated with convergent margins in transition from normal subduction of oceanic crust to collision of a buoyant indentor [1, 12]. This common pattern unites previously disparate tectonic elements in actively deforming subduction/collision regions like the Mediterranean, Papua New Guinea, and New Zealand, among others. In the examples provided here, the timing of buoyant indentor collision, tectonic rotations, and back-arc rifting suggest that the rotations and back-arc rifting are likely to be a consequence of (or are strongly influenced by) the crustal-scale collisional event itself, and are not only due to the widely-embraced mantle-level processes related to slab roll back. However, we do concur with other studies that slab roll back is an important contributor to the process of back-arc rifting and arc curvature.

Our numerical modeling results suggest that the presence of negative stresses (representing slab rollback) at the subduction plate boundary enhances the forearc rotation rate and rates of back-arc rifting (as expected). Although as we show in the model with neutral stresses on the subduction boundary (figure 4) roll back is not a pre-requisite for either forearc rotation or back-arc rifting. We do not suggest that the collision/subduction-induced tectonic rotation model can explain all aspects of back-arc basins and curved plate margins. However, the many natural examples of this process [12] show that the simple model we propose can explain the complex kinematics of deformation at most curved plate margins, particularly at many modern-day active plate margins. These rapid tectonic rotations may even be capable of triggering a localized back-arc rifting event.

6. References

[1] Wallace L M, McCaffrey R, Beavan J and Ellis S 2005 *Geology* **33** 857–60
[2] Wallace L M, Stevens C, Silver E, McCaffrey R, Loratung W, Hasiata S, Stanaway R, Curley R, Rosa R and Taugaloidi J 2004 *J. Geophys. Res.* **109** doi: 10.1029/2003JB002481
[3] Abbott L D, Silver E A, Thompson P R, Filewicz M V, Schneider C and Abdoerrias C 1994 *J. of Sedimentary Res* **B64** 169–83
[4] Taylor B 1979 *Geology* **7** 171-74
[5] Weiler P D and Coe R S 2000 *Tectonophysics* **316** 297–325
[6] Wallace L M, Beavan J, McCaffrey R and Darby D 2004 *J. of Geophys. Res.* **109** doi:10.1029/2004JB003241
[7] Stern T A 1987 *Earth Planetary Sci. Lett.* **85** 265–76
[8] Stern T A, Stratford W R and Salmon M L 2006 *Rev. Geophys.* **44** RG4002 doi:10.1029/2005RG000171
[9] Uyeda S and Kanamori H 1979 *J. Geophys. Res.* **84** 1049–61
[10] Molnar P and Atwater T 1978 *Earth Planetary Sci. Lett.* **41** 330–40
[11] Conrad C P and Lithgow-Bertelloni C 2004 *J. Geophys. Res.* **109** doi:10.1029/2004JB002991
[12] Wallace L M, Ellis S and Mann P submitted Global examples and numerical modelling of the tectonic response to localized collision in subduction settings: Rapid tectonic block rotation, arc curvature, and back-arc rifting *Geochem. Geophys. Geosys.*
[13] McClusky S et al. 2000 *J. Geophys. Res.* **105** 5695–719
[14] Kissel C, Laj C, Poisson A and Gorur N 2003 *Tectonophysics* **362** 199–217
[15] Nyst M and Thatcher W 2004 *J. Geophys. Res.* **109** B11406, doi:10.1029/2003JB002830
[16] Van Hinsbergen D J J, van der Meer D G, Zachariasse W J, Meulenkamp J E 2006 *Int. J. Earth Sci. (Geol. Rundsch.)* doi: 10.1007/s00531-005-0047-5
[17] Reilinger R and 25 other authors 2006 *J. Geophys. Res.* **111** doi: 10.1029/2005JB004051
[18] Martin A 2007 *Tectonophysics* **445** 245–72
[19] Dewey J and Sengor A M C 1979 *Bull. Geol. Soc. Am.* **90** 84–92
[20] LePichon X and Angelier J 1979 *Tectonophysics* **60** 1–42
[21] Karig D E and Kozlu H 1990 *J. Geol. Soc. London* **147** 1023–34
[22] Gautier P, Brun J-P, Moriceau R, Sokoutis D, Martinod J and Jolivet L 1999 *Tectonophysics* **315** 31-72
[23] Sengor A M C 1979 *J. Geol. Soc. London* **136** 269–82
[24] Taymaz T, Jackson J and McKenzie D 1991 *Geophys. J. Int.* **106** 433–90

Post-collisional deformation of the Anatolides and motion of the Arabian indenter: A paleomagnetic analysis

John Piper[1,4], Orhan Tatar[2], Halil Gursoy[2], B Levent Mesci[2], Fikret Kocbulut[2] and Baochun Huang[3]
[1] Geomagnetism Laboratory, Department of Earth and Ocean Sciences, University of Liverpool, Liverpool, L69 7ZE, UK
[2] Department of Geology, Cumhuriyet University, 58140 Sivas, Turkey
[3] Palaeomagnetism and Geochronology Laboratory, Institute of Geology and Geophysics, Chinese Academy of Sciences, Beijing, China

E-mail: sg04@liverpool.ac.uk

Abstract. In the Anatolides of Turkey the neotectonic (post collisional) phase of deformation embraces the period since final closure of the southern arm of Neotethys in mid-Miocene times. The Arabian Shield indenter has continued to deform into the weak Anatolian accretionary collage resulting from subduction of this ocean by a combination of differential movement relative to the African Plate and counterclockwise (CCW) rotation. Much of resulting deformation has been accommodated by slip along major transforms comprising the North Anatolian Fault Zone (NAFZ), the East Anatolian Fault Zone (EAFZ) and the northward extension of the Dead Sea Fault Zone (DSFZ) but has also been distributed as differential block rotations through the zone of weak crust in between. Facets of this deformation comprise crustal thickening and uplift to produce the Anatolian Plateau, establishment of transform faults and tectonic escape as Arabia has continued to impinge into the Anatolian collage. Paleomagnetic analysis of this deformation is facilitated by the widespread distribution of neotectonic volcanism and graben infills, and rotations relative to the Eurasian reference frame are recognised on two scales. Rapid rotation (up to 5°/10,000 years) of small fault blocks is identified between master faults along the intracontinental transforms but deformation does not extend away from these zones and shows that seismogenic upper crust is decoupled from a lower continental lithosphere undergoing continuum deformation. The broad area of weak accreted crust between the transforms is dissected into large fault blocks which exhibit much lower rotation rates (mostly <1°/100,000 years) that vary systematically across the Anatolides. Large CCW rotations near the Arabian indenter diminish westwards to become zero then CW near the limit of tectonic escape in western Turkey. The view that the collage has rotated anticlockwise as a single plate, either uniformly or episodically, during the Neotectonic era is refuted. Instead, deformation has been distributed and differential as the collage has adapted to changing tectonic regimes. Crustal extrusion to the west and south has expanded the curvature of the Tauride Arc and combined with retreat of the Hellenic Arc to produce the extensional horst and graben province in western Turkey. A challenge of present work is to resolve the temporal framework of tectonic rotation. Evidence from the Cappadocian volcanic province and Sivas Basin in central Anatolia indicates that rotation has been concentrated within the last 2-3 million years of the neotectonic era and therefore correlates with establishment of the intracontinental transform framework. Thus we recognise two phases to the evolution of this sector of the orogen: the first embraces crustal thickening and uplift with initiation defined specifically by transition from marine to terrestrial deposition in the Serravallian at ~12 Ma, and the second embraces crustal extrusion to the west motivated by continuing northward movement of Arabia and roll back on the Hellenic Arc since late Pliocene times. Latitudinal motions detected by paleomagnetism are close to confidence limits and consistent with small northward motion of the Anatolides since Eocene times including up to a few hundred km of closure linked to crustal thickening

[4] To whom any correspondence should be addressed.

since the demise of NeoTethys. The driving motion from the Arabian indenter can be partially resolved from the widespread basaltic volcanism that occurred along the periphery of the Arabian Shield at 12-18 Ma during final stages of collision along the Bitlis Suture. This defines CCW rotation of 13-21° with respect to Eurasia. An average CCW rotation of 0.9°/Myr since closure of the Bitlis Suture in mid-Miocene times is unlikely to have been uniform because it has been linked to three adjoining interactions namely episodic opening of the Red Sea, a transition from crustal thickening to tectonic escape in the Anatolian collage and variable rates of strike slip motion on the DSFZ.

1. Scope of Study

In the Middle Eastern sector of the Alpine-Himalayan Orogen closure of the Paleo-Tethys Ocean which accompanied emplacement of the Pontide Orogen was complete by mid-Jurassic times and followed by closure of the Neo-Tethyan Ocean during which terranes were emplaced progressively against the new Eurasian margin prior to final closure along the Bitlis Suture bordering the Arabian Plate in mid-Miocene times.

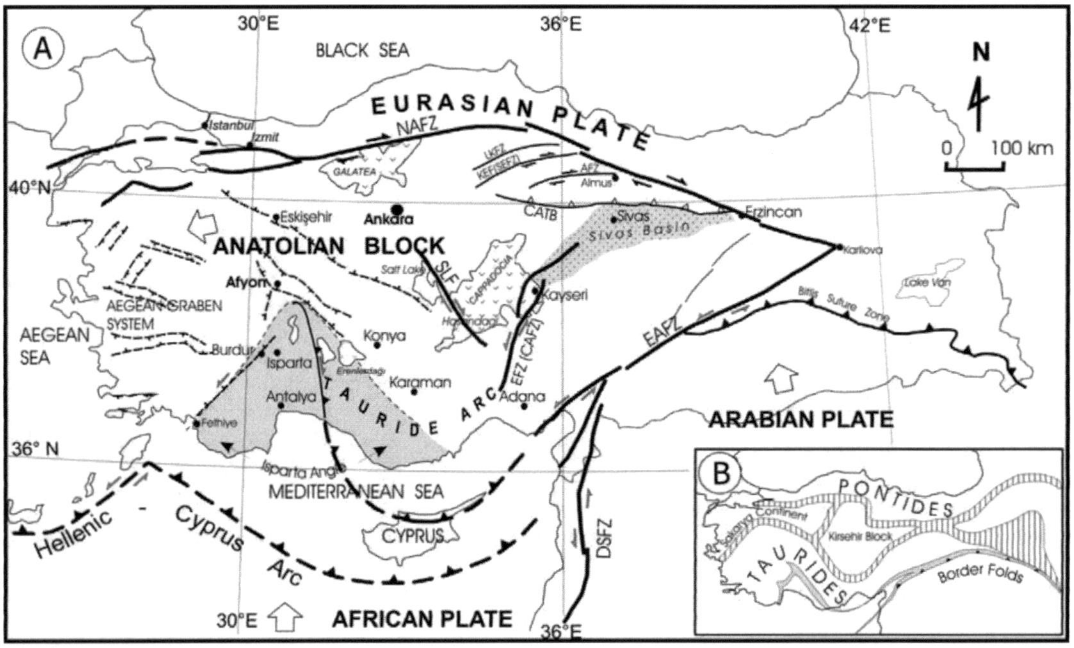

Figure 1. (A) Outline neotectonic map of Anatolia and adjoining regions. The major intracontinental transforms which approximately bound the Anatolia collage accreted from closure of the Neo-Tethys Ocean are abbreviated NAFZ, North Anatolian Fault Zone and EAFZ, East Anatolian Fault Zone; the Dead Sea Fault Zone is abbreviated DSFZ and other neotectonic lineaments are: SLF, Salt Lake Fault Zone; LKF, Laçin-Kızılırmak Fault Zone; KEF (SEF), Kırıkkale-Erbaa Fault Zone (becoming Sungurlu-Ezinepazarı Fault Zone, SEFZ to the west); AF, Almus Fault Zone; EFZ (CAFZ), Ecemiş Fault Zone (Central Anatolian Fault Zone) and CATB, Central Anatolian Thrust Belt which overthrusts the northern margin of the Sivas Basin. The shaded area comprises deformed terranes within the Isparta Angle and was formed mainly during the Paleotectonic era by interference of verging allochthonous units during final stages of Tethyan convergence. **(B)** Framework map showing extent of the main tectonic divisions comprising the Pontide Orogen, the Sakarya and Kirsehir microcontinents and the Tauride Orogen.

Arabia has continued to act as an indenter during differential northward movement and anticlockwise rotation within an interval of neotectonic deformation. Facets of this latter deformation comprise crustal thickening and uplift to produce the Anatolian Plateau, establishment of major intracontinental transforms comprising the North Anatolian Fault Zone (NAFZ) and East Anatolian Fault Zone (EAFZ), extrusion of blocks away from the Arabian Syntaxis by tectonic escape and rapid crustal extension in the Aegean Province in the west (figure 1).

Paleomagnetic study of Miocene volcanic suites emplaced onto the Arabian Platform in southern Turkey shows that this block has been rotating counterclockwise (CCW) at an average rate of 0.9°/Myr since closure of the Bitlis Suture [Authors *in prep.*] which is comparable to the contemporary rate determined from GPS [1]. Slip relative to the African Plate is accommodated by left lateral slip on the DSFZ which merges northwards into a more diffuse fault system at it approaches the EAFZ-Karatas lineament at an unstable FFF junction (figure 1) near Kharaman in southern Turkey. Studies in the bordering Karasu Rift indicate that the bulk of the slip here is occurring on the Amanos Fault Zone at a rate of ~0.45 cm/year suggested by a block model accommodating Brunhes epoch CW rotation [2].

The bulk of the strain imparted by continuing impingement of the Arabian indenter during the neotectonic era has been accommodated largely by deformation of the weak Anatolian accretionary collage between the EAFZ and the NAFZ with the latter now defining the southern margin of the Eurasian Plate (figure 1). This deformation is producing fault block rotations, now widely understood from a compiled database of distributed neotectonic paleomagnetic results, and northward latitudinal movement statistically below levels of paleomagnetic detection [3]. The rotations are on two scales. Very rapid rotations (up to 5°/10,000 years) are identified between master faults of the transform systems notably where extension in pull-apart basins has tapped asthenosphere melts in Brunhes Epoch times such as at Niksar (see figure 2). Rotations involve small fault blocks no more than a few hundred metres in size and do not extend away from these zones. Since rotations do not decline progressively away from the transforms as predicted by continuum models they are evidently confined to the seismogenic upper crust at least partially decoupled from the lower continental lithosphere [4]. Regionally distributed rotations within the Anatolian collage exhibit much lower but variable rotation rates (mostly <1°/100,000 years) which vary systematically across Anatolia: large CCW rotations near the Arabian indenter diminish westwards to become zero near longitude 33°E and then becoming progressively more CW towards the transition zone between compressional and extensional deformation near 29°E. The temporal framework of these rotations is crucial to their tectonic interpretation. Hitherto it has been argued that Anatolia between the NAFZ and EAFZ is a coherent plate rotating CCW either episodically [5] or continuously [6] about an Eulerian pole in the SE Mediterranean region. Such an interpretation would accord with the contemporary GPS record but ignores the ubiquitous CW rotations within the western part of the extruded crustal collage and can only satisfy the paleomagnetic evidence over limited north-south profiles in the central sector. Present evidence for temporal control of rotations indicates that they are variable and mostly very young. Thus rotations observed in rocks emplaced during the Paleotectonic era in northern Turkey are comparable to rotations found in Miocene and younger rocks further south. In the Sivas Basin of central-east Anatolia (figure 1) mean rotations of Paleogene, Miocene and Plio-Pleistocene units are all about 25° CCW whilst further to the north only slightly larger (~35°) CCW rotations are observed in Eocene rocks from both the north and south sides of the NAFZ. The Cappadocian ignimbrite domain of south-central Anatolia (figure 1) includes nine major pyroclastic density current deposits emplaced between 1.1 and 12 Ma in age; after 9 Ma paleomagnetic data from these units define progressive CCW rotation at a rate of ~1.25° [7].

A difficulty with monitoring tectonic rotations is that the volcanics producing the high-fidelity data tend to be either constrained by weak stratigraphic data or have not been studied at a level sufficient to average secular variation; the youngest part of the rotation can potentially be constrained by study of young travertine with ages determine from U-Th series dating [8].

An integrated model for the neotectonic deformation of the Anatolian region must accommodate the progressive change in magnitude and sense of tectonic rotation across Anatolia plus accumulating evidence that they are predominantly very young and probably imparted largely since initiation of the

NAFZ and EAFZ in Late Miocene-Early Pliocene times (figure 2). Although the partitioning of strain across the Anatolian Collage has still to be fully resolved, it is evident that deformation since closure of the Bitlis Suture has been distributed and differential. Indentation and CCW rotation of the Arabian Plate has been responsible for extruding blocks to the west whilst rotating them strongly anticlockwise immediately north of the indenter, and expanding the curvature of the Tauride Arc as the Cyprian Arc has retreated southwards. Sparse Neogene data from eastern Turkey indicate that rotations are CW near the NE margin of the Arabian indenter [Authors *in prep.*].

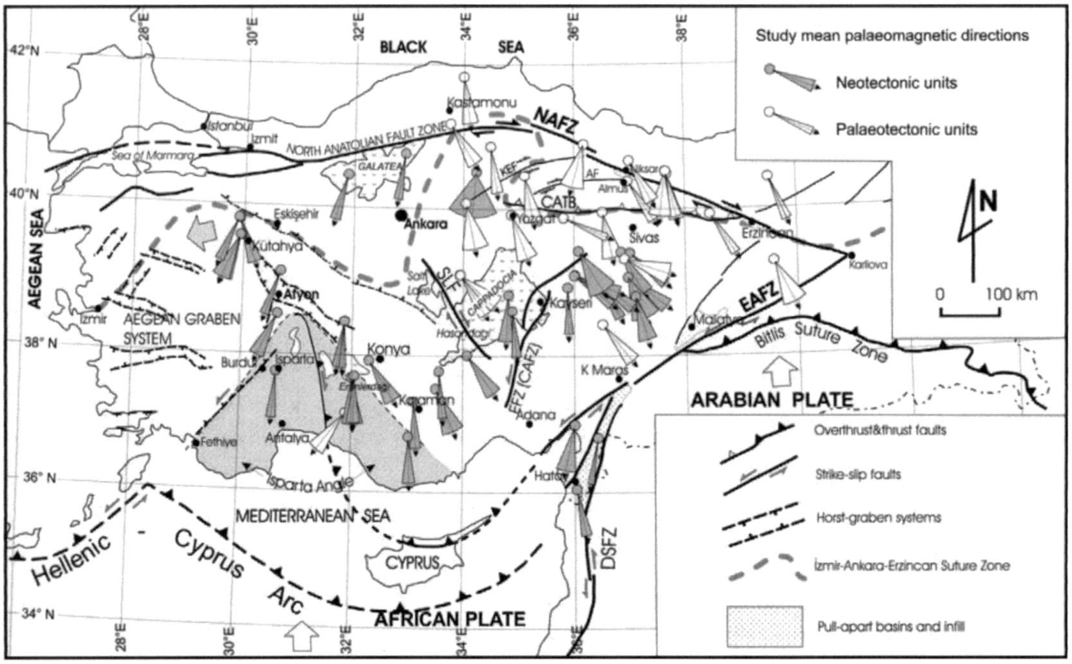

Figure 2. Summary of paleomagnetic directions (all shown as reversed polarity) with 95% confidence arcs from Paleogene and Neogene rock units of Anatolia. Note that, the Sakarya and Kırşehir microcontinents (see figure 1 (B)) were separated by a branch of the NeoTethys and drifted northwards from equatorial latitudes to amalgamate along the İzmir-Ankara-Erzincan Suture Zone shown in this figure in Late Cretaceous and Early Tertiary times; ocean closure commenced during the Late Cretaceous with subduction-related volcanism particularly important in Eocene and Oligocene times as the Kırşehir Block impinged into the Sakarya microcontinent. Other fault lineaments shown in this figure are identified in the caption of figure 1.

Defining the size of blocks involved in these differential rotations is a challenge to current work. In the Afyon region near the western margin of the extruded blocks (figure 1) consistent 10-20° CW rotation is identified across a crustal segment at least 120 km in N-S extent and 100 km wide [9] whilst palaeomagnetic studies of the Kepezdag and Yamadag volcanic complexes near the southern margin of the Sivas Basin identify coherent ~25° CCW rotation across an area at least 150 km N-S and 120 km E-W [Authors *in prep.*]. Thus although block boundaries are typically eroded zones and concealed by Neogene infill, it is apparent that some at least are large and likely to be defined by subsidiary strike slip faults such as those splaying SSW from the NAZF at an acute angle to the maximum principal stress (σ_1) in northern Anatolia (figure 1). North of the Almus Fault (figure 1) for example, paleomagnetic study has already noted larger CCW rotation in contemporaneous Eocene volcanics on the south side of the NAFZ compared with those on the north side [10].

Inclination shallowing tends to be only a subsidiary effect in the paleomagnetic database from Anatolia and is insignificant in many results from SE Turkey. This is probably because the source material is mostly igneous in origin. Nevertheless mean inclinations derived from latitudinal sectors across Anatolia yield paleolatitudes a few degrees shallower than predicted values and since these results include very young material they cannot be accommodated by northward movement during ongoing collision. There remains a residual difference between mean Eocene, Miocene and Plio-Pleistocene results equivalent to up to 300km which could represent the effect of compression and uplift [3]. This is likely to have been most important immediately north of the indenter where strong CCW rotation is found (figure 2); further west tectonic escape may have replaced crustal thickening by compression whilst in the extensional domain of western Turkey degrees of stretching of 30-50% have been proposed and southward motion is likely. The most important geological constraint is the environmental transition from marine to terrestrial conditions at ~11.8 Ma during late Serravallian-earliest Tortonian times recording initial uplift of the Anatolian Plateau. Only some 8 Myr later with establishment of the intracontinental transforms does strain accommodation by tectonic escape seem to have taken over [3].

The present velocity of the crust in central Anatolia deduced from GPS is ~2 cms/year to the WSW relative to Eurasia suggesting CCW rotation about a pole in the vicinity of the Nile Delta equivalent to a rate of $1.3 \pm 0.1°$/Ma [1]. The rates deduced from palaeomagnetism are both variable and up to an order higher than this with the exception of the central part of the Tauride Arc where rotation has been near zero. In contrast the rotation rate of the rigid Arabian Block since mid-Miocene is statistically the same as that deduced from GPS. Thus the short-term assessment derived from GPS seems to be of much longer term application when applied to rigid plates but fails to predict long term deformation within weak crust like the Anatolian accretionary collage. The entire spectrum of techniques used for the analysis of neotectonism is therefore relevant to unraveling the strain accommodation within such regions.

2. References

[1] McClusky S et al. 2000 *J. Geophys. Res.* **105** 5695–719
[2] Tatar O, Piper J D A, Gürsoy H, Heimann A and Koçbulut F 2004 *Tectonophysics* **385** 17–43
[3] Piper J D A, Tatar O, Gürsoy H, Koçbulut F and Mesci B L 2006 *GSA Bull. Spec. Paper* **409** 417–39
[4] Piper J D A, Tatar O and Gürsoy H 1997 *Earth Planet. Sci. Lett.* **150** 191–203
[5] Platzman E S, Tapirdamaz C and Sanver M 1998 *Tectonophysics* **299** 175–89
[6] Kissel C, Laj C, Poisson A and Görür N 2003 *Tectonophysics* **362** 199–217
[7] Piper J D A, Gürsoy H and Tatar O 2003 *J. Volcanol. Geoth. Res.* **117** 237–62
[8] Piper J D A, Mesci L B, Gürsoy H, Tatar O and Davies C J 2007 *Phys. Earth Planet. In.* **16** 50–73
[9] Gursöy H, Piper J D A and Tatar O 2003 *Tectonophysics* **374** 57–79
[10] Tatar O, Piper J D A, Park R G and Gürsoy H 1995 *Tectonophysics* **244** 251–66

IOP Conf. Series: Earth and Environmental Science **2** (2008) 012012 doi:10.1088/1755-1307/2/1/012012

Metamorphic and Ar/Ar geochronology constraints on the Alakeçi shear zone: Implications for the extensional exhumation history of the northern Kazdağ Massif, NW Turkey

Nikolay Bonev[1,5], **Laurent Beccaletto**[2], **Martin Robyr**[3] **and Patrick Monié**[4]
[1] Sofia University "St. Kliment Ohridski", Department of Geology and Paleontology, Sofia 1504, Bulgaria
[2] BRGM Geological Survey/Geology of Sedimentary Basins, 45060 Orléans, Cedex 2, France
[3] Geological Sciences Department, University of Texas, Austin, TX 787 12-0254, USA
[4] Géosciences, University of Montpellier 2, UMR 5243, 34095 Montpellier Cedex 5, France

E-mail: niki@gea.uni-sofia.bg

Abstract. The Kazdağ Massif exposes a metamorphic dome in the Biga Peninsula of northwest Turkey. An extensional origin has been proposed for the dome, limited on both flanks by detachments and/or shear zones. The northern flank is bounded by the extensional Alakeçi Shear Zone (ASZ), whose P-T-t path is still poorly known. We therefore focus on its metamorphic conditions and related temporal history to precise its tectono-metamorphic evolution. The local tectonostratigraphy in structurally ascending order comprises: (i) the high-grade metamorphic core rocks of the Kazdağ Massif (gneisses and micaschists intercalated with amphibolites and marbles); (ii) the two kilometer-thick ASZ; (iii) the overlaying unmetamorphosed pre-Cenomanian accretionary Çetmi mélange; and (iv) Neogene sedimentary and volcanic cover rocks. ASZ mylonites were derived from both the core rocks and the mélange lithologies. From the north to the south the mylonitic fabrics in the ASZ depict a top-to-the N-NNE shearing, parallel to the NNE-plunging stretching lineation and NNW-dipping mylonitic foliation. This geometry implies normal sense movement i.e. north-side down-dip extensional displacement along this flank of the Kazdağ Massif. The northward transition from ductile to brittle-ductile regime through the ASZ shows that the deformation occurred at decreasing temperatures and degree of metamorphism. The paragenesis in equilibrium within the mylonitic gneisses and schists contains Qtz + Fs + Ms + Bt + Grt ± St ± Sill, with late retrogressive chlorite after biotite and garnet. Four samples of ASZ rocks yielded pressures between 6.9-5.7 kbar and temperatures between 706-587 °C. Three samples from the mylonitic rocks supplied in situ isochron $^{36}Ar/^{40}Ar$ mica ages between 31.2-24.2 Ma, which we interpret to date the cooling of the mylonites following the P-T decrease across the ASZ. The metamorphic and structural results support the extensional character of the ASZ, and sketch transition from sillimanite core gneisses in the deeper structural levels to chlorite schists towards the top of the shear zone. These new data allow to precise the peak P-T conditions and the temporal evolution in the northern flank of the Kazdağ Massif, where Late Oligocene

[5] To whom any correspondence should be addressed.

extensional exhumation was assisted by NNE-directed ductile-brittle ASZ, which had operated from amphibolite to greenschist facies. At the regional scale, this tectono-metamorphic pattern is similar to those observed on other places of the north Aegean domain.

IOP Conf. Series: Earth and Environmental Science 2 (2008) 012013 doi:10.1088/1755-1307/2/1/012013

Monazite geochronology, magmatism, and extensional dynamics within the Menderes Massif, western Turkey

Elizabeth J Catlos[1,2,5], Courteney B Baker[2], Sorena S Sorensen[3], Ibrahim Çemen[2] and Mete Hancer[4]

[1] University of Texas at Austin, Jackson School of Geosciences, Geological Science Dept., 1 University Station C1100, Austin, TX 78712-0254, USA

[2] Oklahoma State University, School of Geology, 105 Noble Research Center, Stillwater, OK 74078, USA

[3] Smithsonian Institution Museum of Natural History, PO Box 37012, MRC 119, Washington, DC, 20013-7012, USA

[4] Pamukkale Universitesi, Muhendislik Fakultesi, Jeoloji Muh. Bolmu, Denizli, 20070 Turkey

E-mail: ejcatlos@gmail.com

Abstract. Geochemical and geochronological data were collected from S-type, peraluminous granites (Salihli and Turgutlu) that intrude a detachment that bounds the northern edge of the central Menderes Massif core complex (Aegean region, western Turkey). The granites may have been generated due to subduction of the Eastern Mediterranean floor along the Hellenic trench. In situ Th-Pb ion microprobe monazite ages from the rocks range from 21.7±4.5 Ma to 9.6±1.6 Ma (±1σ), which could record their exhumation history. Higher uncertainty in the ages is attributed to monazite common Pb, but the range is consistent with cathodoluminescence (CL) images that document complex textures within the granites. Salihli and Turgutlu granites share many similar characteristics, including multiple generations of plagioclase, plagioclase replacing K-feldspar and the development of myrmekite, evidence for fluid interaction, and multiple generations of microcracks. Ages reported here are similar to dates constraining extension reported elsewhere in the Aegean, but indicate additional complexities when linking movement within the Menderes Massif to large-scale geodynamic processes that created other metamorphic core complexes in the region. Difficulties exist in linking the ages obtained from the granites to specific tectonic events due to the presence of secondary alteration textures, generations of mineral growth and multiple episodes of deformation.

1. Introduction

The Menderes Massif in southwestern Turkey is a metamorphic core complex that exposes amphibolite-facies rocks with ages of prograde metamorphism ranging from the Cambro-Ordovician to the Eocene-Oligocene (figure 1) [see review by 1]. Extensional exhumation began sometime during the mid- to late Cenozoic, although the specific timing has been debated [e.g., 1, 2, 3]. The massif is thought to be one of the largest metamorphic core complexes in the world [3], exposing ~ 40,000 km^2 of metamorphic and igneous rocks.

The Menderes Massif is only one of several domains in the Aegean region that experienced a

[1] To whom any correspondence should be addressed.

switch from compressional to extensional tectonics, notably the Rhodope, Kazdag, Cycladic, and Crete Massifs [e.g., 4, 5, 6, 7, 8]. The similarities among the massifs have been explored [e.g., 9. 10, 11] and further advances require integration of metamorphic, geochronologic, and stratigraphic data.

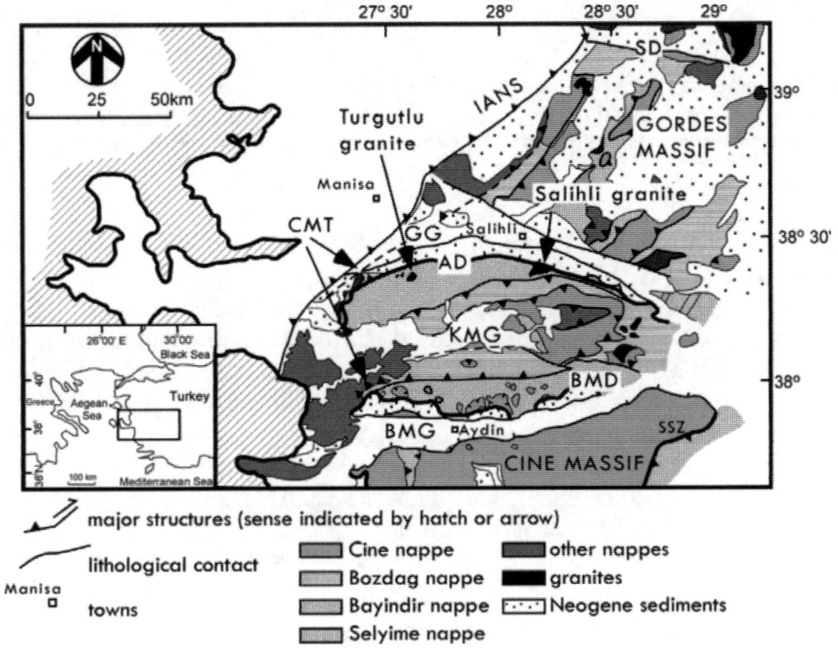

Figure 1. Map of the Menderes Massif [after 12 and 13]. Insets show the location of the massif in relation to western Turkey. Abbreviations (from north to south): SD= Simav Detachment; CMT= Cyclades Menderes Thrust, GG= Gediz Graben; AD= Alasehir Detachment; KMG= Küçük Menderes Graben; BMD= Büyük Menderes Detachment; BMG= Büyük Menderes Graben.

Here we report Th-Pb ion microprobe monazite (a rare-earth phosphate) ages and cathodoluminescence (CL) images from two synextensional granites (Salihli and Turgutlu) located ~50 km apart along the northern Alasehir (Kuzey) detachment of the Menderes Massif (figure 1). Monazite remains impervious to diffusive Pb loss at conditions >900°C [14] and appears resilient to radiation damage [15]. Fluid-mediated dissolution-reprecipitation reactions that occur along a retrograde path can affect monazite ages [e.g., 16]. Fluids are often invoked to trigger of reactions of monazite growth [17, 18]. CL has been used to document mineral zoning, deformation, and fluid alteration a variety of lithologies [e.g., 19, 20, 21]. CL is often used as a guide for dating of specific zones within zircon [e.g., 22, 23]. Monazite does not luminesce, but CL examinations of granites have documented magmatic mineral features overprinted by alteration, and may be "necessary in order to obtain valid information about the origin of granites" [19].

2. Geologic Background

This study focuses on the central Menderes Massif's north-dipping Alasehir detachment (figure 1). The structure is located along the northern margin of central Menderes metamorphic core complex and exposes garnet-bearing schists and gneisses of the Bayindir and Bozdag nappes intruded by the Salihli and Turgutlu granites [24, 25]. The detachment dips ~10-20°N, is exposed ~150 km along strike, and plays a key role in developing models for the evolution of the massif [e.g., 1, 13, 24]. Higher angle normal faults cut the structure and separate the metamorphic assemblages from sedimentary units which bound the southern edge of the Gediz graben [e.g., 26].

The Alasehir detachment may have originally initiated at high angle (48-53°) [27] and rolled back to its present-day low angle [12, 13, 24]. Several researchers link activity along the Alasehir detachment to the Büyük Menderes detachment, which bounds the southern edge of the central Menderes Massif [e.g., 13], and to the Simav detachment, located in the Gordes Massif (figure 1) [e.g., 28, 29]. The central Menderes metamorphic core complex may have exhumed due to bivergent motion along both the Alasehir and Büyük Menderes detachments [13].

Granites that intrude the Alasehir detachment are separated by an E-W distance of ~50 km (figure 1), and their petrology and structural setting suggest syntectonic emplacement during extension and low- to mid-greenschist facies conditions during ductile deformation [24, 25, 28]. The Salihli samples we analyzed contain Qtz + Pl + Kfs + Bt + Ap + Zrn ± Ilm ± Chl ± Ms ± Px ± Aln ± Ep ± Xtm ± Mnz ± Thorite ± Ttn ± Rt ± Hem ± Cld; abbreviations after [30]. Only one Salihli sample (CC20) contains monazite. Turgutlu samples have a similar mineral assemblage, but do not contain Px, Ep, Th, or Cld.

Monazite grains were dated in rock thin section using the Th-Pb ion microprobe method [31]. $^{40}Ar/^{39}Ar$ ages of 13.1±0.2 Ma from Turgutlu and 12.2±0.4 Ma from Salihli biotite grains indicate extension was already active by the Late Miocene [24, 25]. Recently, 16.1±0.2 Ma and 15.0±0.3 Ma U-Pb ages were reported from aliquots of monazite and allanite grains separated from the granites [25]. Turgutlu allanite grains in the samples we collected are anhedral with diffuse boundaries, filling cracks or crystal faces within other minerals, typically K-feldspar [32]. In contrast, Salihli allanite grains are large (mm-sized), zoned, and can be significantly affected by radiation damage and microcracks [32]. These grains often contain inclusions of other minerals (Zrn ± Ap ± Qtz). Biotite $^{40}Ar/^{39}Ar$ ages from an unspecified granite located on the Alasehir detachment range from 36.4±2.2 Ma to 16.6±0.3 Ma [33], the ~20 m.y. difference likely due to the presence of excess ^{40}Ar.

3. Cathodoluminescence images and monazite ages

Monazite ages from Salihli sample CC20 range from 21.7±4.5 Ma to 9.6±1.6 Ma (±1σ), whereas those from four Turgutlu granites range from 19.2±5.1 Ma to 11.5±0.8 Ma. The average of all of the monazites dated is 15.0±2.8 Ma (MSWD= 3.8). The level of uncertainty reported for these Menderes Massif monazite grains is higher than is typically reported, and is attributed to lower than average radiogenic $^{208}Pb^*$ contents. The average $^{208}Pb^*$ for the dated grains is 71±4%. Figure 2 shows CL images from sample CC20, which is the only Salihli granite that contains monazite. Figure 3 shows CL images from samples of the Turgutlu granite.

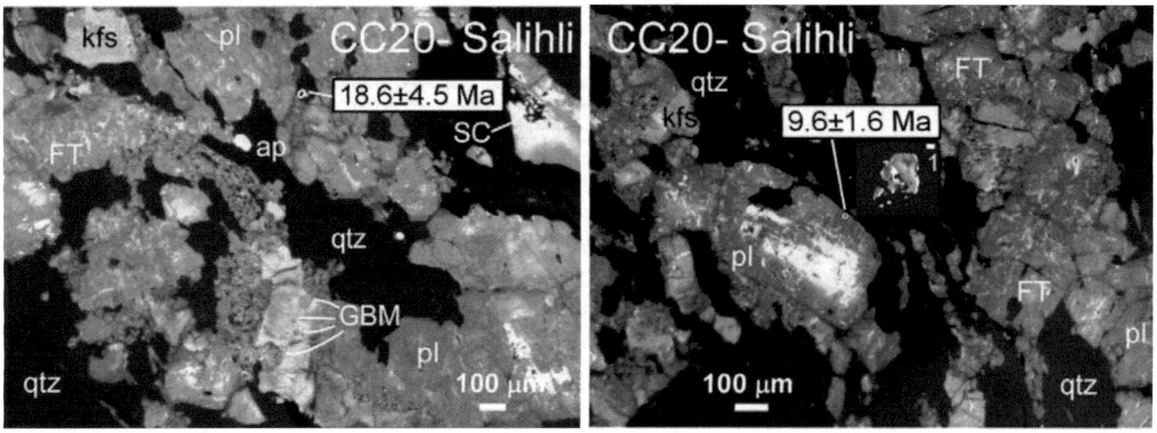

Figure 2. CL images of Salihli granite CC20. Monazite grains outlined and ages (±1σ) indicated. Inset in the right image shows a backscattered electron image (BSE) image of the dated grain. Mineral abbreviations after [30]. Other abbreviations: GBM= grain boundary migration; FT= flame texture; SC= shattered core.

We identify five shades of blue displayed by the K-feldspar grains in the Salihli sample (figure 2). The cause of blue in K-feldspar has been discussed by a number of researchers [e.g., 34, 35, 36], and may be related to compositional differences and/or alteration. Many K-feldspar grains in both granites are texturally resorbed and are significantly affected by grain boundary migration and recrystallization, as yellow or brown plagioclase grains bulge into the blue K-feldspar.

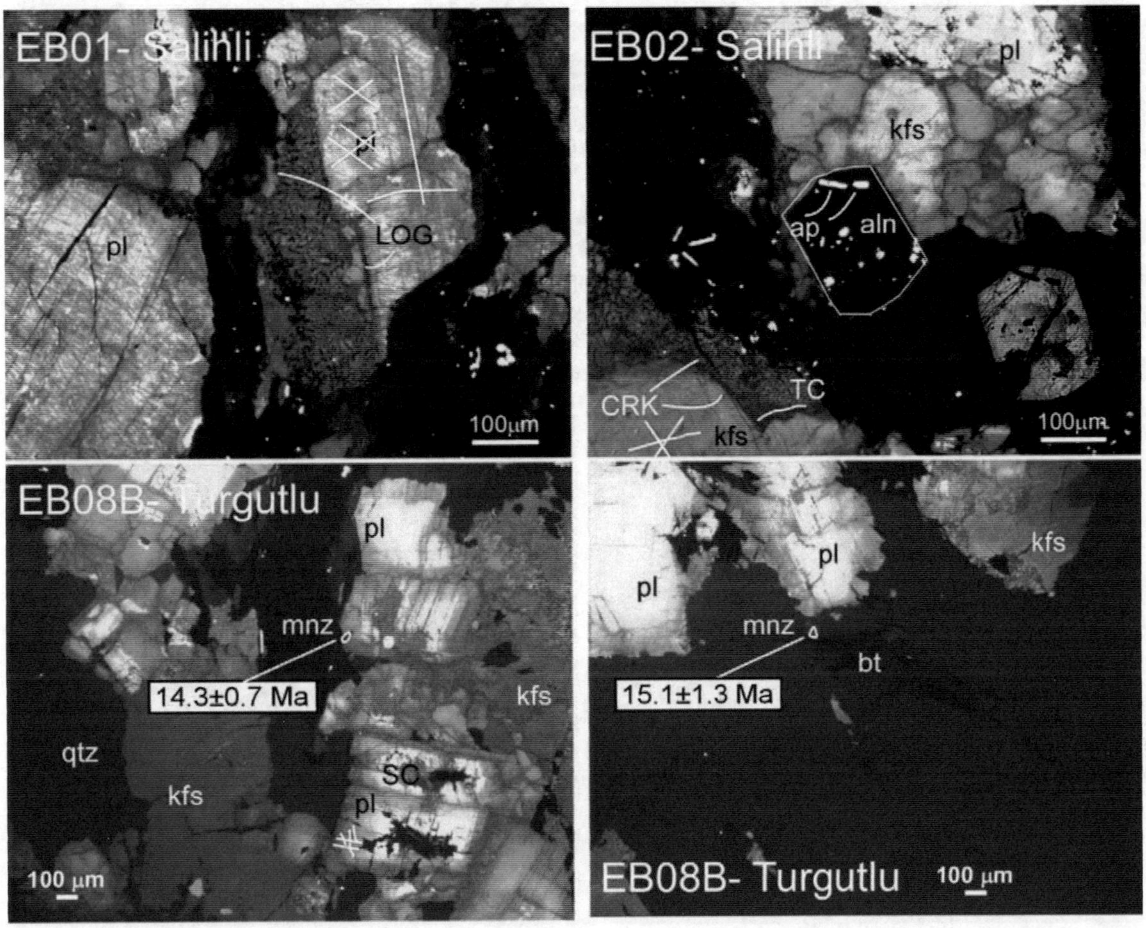

Figure 3. CL images of Salihli samples EB01 (upper left) and EB02 (upper right) and Turgutlu samples EB08B (lower left and right). Inset in the upper right image shows a BSE image of the outlined allanite grain. Mineral abbreviations after [30]. Other abbreviations are indicated to draw attention to some key features: LOG= left over grain; CRK= crack; TC= terminating cracks; SC= shattered core. Some cracks are outlined in white.

Many K-feldspar grains in the granites are replaced by myrmekite and boundaries between grains are red-brown (figures 2 and 3). The change of color in K-feldspar from blue to red is linked to fluid interaction [35], and we suspect fluids infiltrated cracks between grains. The source of the fluids is unknown, but evidence exists for both post-crystallization hydrothermal fluids (as cracks appear filled by material) and/or late-stage magmatic fluids (as driving the development of myrmekite). Plagioclase grains in the granites are shades of yellow and brown (figure 2) or green and olive (figures 3-5). Some plagioclase grains show bright "flame-type" structures (figure 2). Incomplete elimination of K-feldspar led to the preservation of "left-over" or "relict" grains (figure 3). The cores of some plagioclase grains in the granites contain bright CL domains that are corroded and cracked (figures 2 and 3). [37] report similar plagioclase textures and suggested the cores are "shattered", relics of crystals from a previous magma, and evidence for magma mixing.

Multiple generations of microcracks are present in the granites. In general, thicker (~10-100 μm-wide) black cracks within bright plagioclase cores are restricted to the core (figures 2 and 3). Three sets of microcracks are seen as narrow (~1-2μm-size) lines in plagioclase and K-feldspar, and can occur in swarms, generally at 90° or 120° angles from each other. These cracks are overprinted by (1) wider (5-10μm), roughly linear intragranular cracks generated due to subsequent deformation and accommodated by the opening of existing microcracks, (2) irregular, stepped, intragranular cracks that cross-cut plagioclase grains to connect cracks, and (3) transgranular cracks that cut the rock and link fractures. Some intragranular cracks extend from plagioclase grain rims inwards, whereas others terminate within a grain and fail to extend to the outer rim (figure 3).

Monazite is present in the Salihli and Turgutlu granites as matrix grains and as inclusions in rims of plagioclase or K-feldspar. The textural relationships suggest monazite grew late in the crystallization sequence. The youngest monazite (9.6±1.5 Ma) in the Salihli sample is included in the outermost rim of a plagioclase grain (figure 2). A high-contrast BSE image shows that this monazite is ahedral and is patchily zoned. We suspect this young monazite has been affected by retrogression.

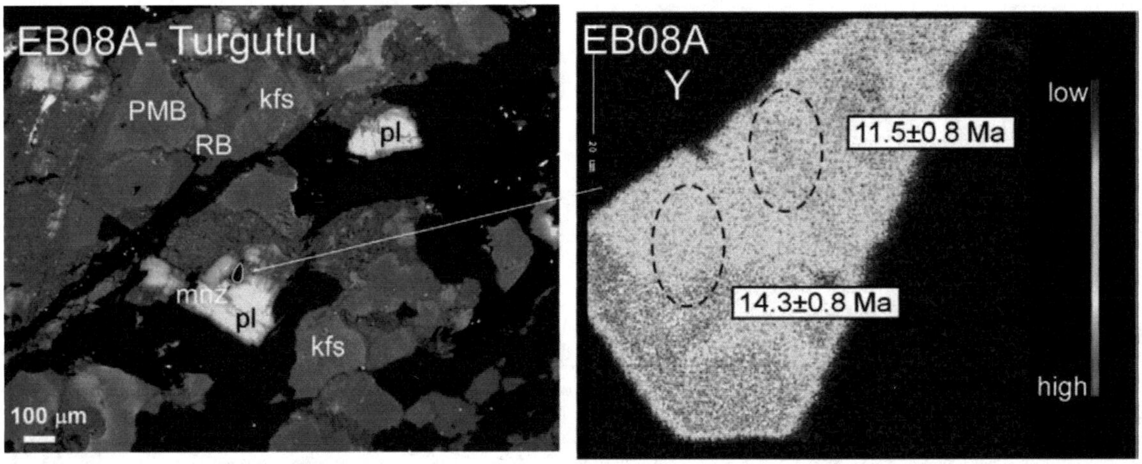

Figure 4. CL image of Turgutlu samples EB08B (left) and X-ray Y map of the dated monazite. Mineral abbreviations after [30]. Other abbreviations: PMB= patchy medium blue; RB= reddish blue.

In Turgutlu sample EB08A, a triangular shaped monazite is included in the rim of a plagioclase grain (figure 4). One edge of the monazite shows lower Y (and U contents, see [32]) and is located near a region where myrmekite is growing in contact with the plagioclase. The edge of the monazite closer to the myrmekite is 11.5±0.8 Ma, whereas the region closer to the core of the plagioclase is 14.3±0.8 Ma. We suggest the younger age is the result of monazite resetting due to fluid infiltration associated with myrmekite formation. The older age is similar to the ages of monazite inclusions in plagioclase rims in other Turgutlu samples (14.9±1.2 Ma; 14.3±0.7 Ma; 13.8±0.9 Ma) (e.g., figure 3) and could time a crystallization episode of plagioclase in these rocks.

4. Discussion and conclusions

Monazite ages from the Salihli and Turgutlu granites range from 21.7±4.5 Ma to 9.6±1.6 Ma (±1σ); the ~10 m.y. difference is consistent with granites' complex textures. These granites experienced multiple episodes of mineral growth and brittle deformation. The CL images show evidence of grain boundary migration and recrystallization between plagioclase and K-feldspar. The presence of myrmekite and relict grains indicates that plagioclase replaces K-feldspar in these rocks due to the infiltration of fluids. Generations of microcracks document that these rocks experienced deformation at a range of temperatures. An initial generation of plagioclase grains shattered at higher temperatures possibly due to magma mixing, and subsequently, the rocks experienced episodes of brittle deformation at lower temperatures. Sediment data are consistent with at least two episodes of extension along the Alasehir detachment [26, 38]. [38] suggests an intervening short-lived period of compression, but this has been debated [39].

The Salihli and Turgutlu granites may have formed under a compressional regime due to the north-dipping subduction of the Eastern Mediterranean floor along the Hellenic trench. During subduction, a generation of plagioclase crystals and K-feldspar crystallized in the rocks. We suggest that the shattered K-feldspar grains are evidence of the initial generation. As we do not have monazite grains located in the cores of these grains, we cannot constrain the timing of this event. The oldest ages we obtained are the most uncertain (21.7±4.5 Ma; 18.6±4.5 Ma; 18.3±1.8 Ma Salihli and 19.2±5.1 Ma Turgutlu). We speculate that the ~14-15 Ma ages recorded by these granites (figures 3 and 4) [25] time crystallization of monazite as the granite experienced continued growth of plagioclase and K-feldspar. We speculate that this is the main phase of crystallization of plagioclase and K-feldspar in these granites. Of interest are the youngest monazite ages, which, in the Turgutlu granite, is located in the rim of a K-feldspar grain close to myrmekite (11.5±0.8 Ma), and in the Salihli granite, is located in the outer edge of an altered plagioclase crystal (9.6±1.6 Ma). These ages may time an episode of deformation as the granites were subjected to extension and exhumation along the Alasehir detachment [e.g., 40].

The monazite ages reported here are similar to dates thought to constrain extension reported elsewhere in the Aegean. For example, rocks in the Rhodope Massif record early Tertiary [e.g., 41] to the mid-Miocene extension [e.g., 42]. The Kazdag massif forms a metamorphic core complex of Oligo-Miocene age and experienced post-Miocene detachment faulting [e.g., 5]. Granites from the island of Naxos, Greece contain Early to Mid-Miocene zircons [4]. Lithologies in these regions may not correlate [e.g., 10], but the timing of large-scale extension may overlap. In addition, many of these domains in the Aegean contain granites cut by large-scale detachments. CL imagery is a useful tool for comparing these rocks. Mineral textural relationships, growth histories, and deformation structures are clearly identifiable. The CL images of granites in the Menderes Massif reported here can be compared to granites that are in a similar tectonic relationship in the Cyclades, for example. In situ monazite ages, placed in a textural context, are key for deciphering the complex set of processes that worked to exhume the Aegean metamorphic core complexes.

Acknowledgments
This work was supported by an award to E.J Catlos and I. Çemen from the National Science Foundation (NSF- 0440169). We thank the UCLA Ion Microprobe Facility, which is partly supported by a grant from the Instrumentation and Facilities Program, Division of Earth Sciences, National Science Foundation. CC and EB samples were collected with assistance from M.J. Kohn, M. Cemal Göncüoglu, and E. Diniz. We appreciate the comments by M. J. Kohn that improved the original version.

5. References
[1] Çemen I, Catlos E J, Gogus O and Ozerdem C 2006 *GSA Bull. Special Paper* **409** 353–79
[2] Yılmaz Y, Genç S C, Gűrer F, Bozcu M, Yılmaz K, Karacik Z, Altunkaynak S and Elmas A 2000 *Geol. Soc. London Special Pub.* 173 353–84
[3] Westaway R 2006 *Tectonophysics* **412** 1–25
[4] Keay S, Lister G and Buick I 2001 *Tectonophysics* **342** 275–312
[5] Duru M, Pehlivan S, Senturk Y, Yavas F and Kar H 2004 *Turkish J. Earth Sci.* **13** 177–86
[6] Bonev N 2006 *GSA Bull. Special Paper* **409** 211–35
[7] Brichau S, Ring U, Ketcham R A, Carter A, Stockli D and Brunel M 2006 *Earth Panet. Sci. Lett.* **241** 293–306.
[8] Chatzaras V, Xypolias P and Doutsos T 2006 *Geol. Mag.* **143** 859–76

[9] Papanikolaou D J and Demirtasli E 1987 Geological correlations between the Alpine segments of the Hellenides-Balkanides and Taurides-Pontides. *Pre-Variscan and Variscan events in the Alpine-Mediterranean Mountain Belts*, eds H W Flügel, F P Sassi, P Grecula (Bratislava:Alfa Publishers) pp 387–96
[10] Ring U, Gessner K, Güngör T and Passchier C W 1999 *J. Geol. Soc. London* **156** 3–6
[11] Jolivet L, Rimmele G, Oberhaensli R, Goffe B and Candan O 2004 *Bull. Soc. Geol. France* **175** 217–38
[12] Sözbilir, H 2001 *Turkish J. Earth Sci.* **10** 51–67
[13] Gessner K, Ring U, Passchier C W, Johnson C, Hetzel R and Güngör T 2001 *Geology* **29** 611–14
[14] Cherniak D J, Watson E B, Grove M and Harrison T M 2004 *Geochim. Cosmochim. Acta* **68** 829–40
[15] Meldrum A, Boatner L A, Weber W J and Ewing R C 1998 *Geochim. Cosmochim. Acta* **62** 2509–20
[16] Seydoux-Guillaume A-M, Paquette J-L, Wiedenbeck M, Montel J-M and Heinrich W 2002 *Chem. Geol.* **191** 165–81
[17] Harlov D E and Förster H-J 2003 *Am. Min.* **88** 1209–99
[18] Harlov D E, Wirth R and Förster H-J 2005 *Contrib. Mineral. Petrol.* **150** 268–86
[19] Ramseyer K, AlDahan A A, Collini B and Landstrom O 1992 *Tectonophysics* **216** 195–204
[20] Sorensen S, Harlow G and Rumble D 2006 *Am. Min.* **91** 979–96
[21] Pidgeon R T, Nemchin A A, van Bronswijk W, Geisler T, Meyer C, Compston W and Williams I S 2007 *Geochim. Cosmochim. Acta* **71** 1370–81
[22] Aleinikoff J N, Wintsch R P, Fanning C M and Dorais, M J 2002 *Chem. Geol.* **188** 125–47
[23] Hanchar J M and Miller C F 1993 *Chem. Geol.* **110** 1–13
[24] Hetzel R, Ring U, Akal C and Troesch M 1995 *J. Geol. Soc. London* **152** 639–54
[25] Glodny J and Hetzel R 2007 *Geol. Mag.* **144** 235–46
[26] Purvis M and Robinson A 2005 *Sedimentary Geol.* **173** 373–408
[27] Bozkurt E and Sözbilir H 2004 *Geol. Mag.* **141** 63–79
[28] Isık V, Seyitoğlu G and Çemen I 2003 *Tectonophysics* 374 1–18
[29] Ring U and Collins A S 2005 *J. Geol. Soc. London* **162** 289–98
[30] Kretz R 1983 *Am. Min.* **68** 277–9
[31] Harrison T M, McKeegan K D and Le Fort P 1995 *Earth Planet. Sci. Lett.* **133** 271–82
[32] Catlos E J, Baker C B, Çemen I and Ozerdem C 2008 *Mineralogia* **38** 5–18
[33] Delaloye M and Bingöl E 2000 *Int. Geol. Rev.* **42** 241–68
[34] Marshall D J 1988 *Cathodoluminescence of geological materials* (Unwin Hyman Publishers, London)
[35] Finch A A and Klein J 1999 *Contrib. Mineral. Petrol.* **135** 234–43
[36] Parsons I, Steele D A, Lee M R and Magee C W 2008 *Am. Min.* **93** 875–79
[37] Janoušek V, Braithwaite C J R, Bowes D R and Gerdes A 2004 *Lithos* **78** 67–99
[38] Koçyiğit A, Yusufoglu H and Bozkurt E 1999. *J. Geol. Soc. London* **156** 605–16
[39] Seyitoğlu G, Çemen I and Tekeli O 2000 *J. Geol. Soc. London* **157** 1097–100
[40] Yilmaz Y, Genc, C, Karacik Z and Altunkaynak S 2001 *J. Geodyn.* **31** 243–71
[41] Beccaletto L, Bonev N, Bosch D and Bruguier O 2007 *Geol. Mag.* **144** 393–400
[42] Dinter D A, Macfarlane A, Hames W, Isachsen C, Bowring S and Royden L 1995 *Tectonics* **14** 886–908

Nature of Beypazari Granitoid: Geology and geochemistry, Northwest Anatolia, Turkey

Yusuf Kagan Kadıoğlu[1] **and Oguz Zoroğlu**
Ankara University, Engineering Faculty, Geological Engineering Department, Ankara, Turkey

E-mail: kadi@eng.ankara.edu.tr

Abstract. Beypazari Granitoid represents one of the widest exposure bodies within the Sakarya continent. It has several exposure bodies at Beypazari, Oymaagac, Tahir, Kirbasi and Yalnizcam of Eskisehir city. Although Beypazari Granitoid has several independence outcrops, the field geology and the aeromagnetic anomaly reveal that they are a unique body at the lower part of the crust. Field observations, mineralogical and petrographical investigations reveal that Beypazari Granitoid is composed of five different units these are monzonite, quartz monzonite, granodiorite, granite and alkali feldspar granite. Alkali feldspar granite cuts all the other subunits in the form of aplitic dyke. All the other subunits have gradual contact with the each other. Excluding alkali feldspar granite mafic enclaves are observed within the other 4 subunits as angular and elliptical in shape and changing from mm up to 20 cm in size. These enclaves can be divided genetically into three different types according to the field observation, textural features and mineralogical compositions. The first type has igneous texture, sharp contact with host rock, rimmed by fine crystalline mafic minerals and represents the abundant enclave type within the monzonite, quartz monzonite, granodiorite and granite as elliptical to sub-angular in shape. These enclaves mostly have subophitic texture in the composition of diorite, quartz diorite and monzodiorite. Ocellar quartz, acicular apatite, poiclitic feldspars and blade shaped biotite are the most characteristic features of the first type of these enclaves, which may represent the magma mixing/mingling enclaves in origin. The second type of these enclaves has a cumulate texture and is representing a segregation of early crystallization processes of mafic minerals. The second type of these enclaves is the product of the early crystallization of magmatic differentiation and is forming the magma segregation enclave types. The third type of the enclaves have metamorphic texture with clear metamorphic lineation, sharp contact with host rock and mostly observed at the northwest part of Kirbasi and Tahir region in the form of xenolithic enclaves. They have angular to sub-angular in shape. These types of the enclaves have hornfels in composition at the contact with the host rock as a product of contact metamorphism and amphibolites in composition at the core as a product of high temperature and middle pressure metamorphism. The textural features and mineral composition of the third type of the enclaves may indicate a fragment of metapelitic rocks, which caught by the granitoid magma in the form of xenolithic enclaves. Whole rock geochemistry reveals that Beypazari granitoids are subalkaline and calcalkaline in nature. They are enriched in Light-REE and LIL with respect to High-REE and HFS elements. Tectonic discrimination diagrams of Beypazari granitoid suggest a product of plate convergence and probably belong to Volcanic Arc Granitoid (VAG). The field observations, mineralogy, petrography with the whole geochemical data reveal that the Beypazari Granitoid magmas are derived from a subduction-modified magma and metasomatized mantle source with considerable crustal contribution.

[1] To whom any correspondence should be addressed.

1. Introduction

The Beypazari Granitoid (BG) in Turkey is a felsic intrusive of Sakarya continent that played a significant role during the Tethyan evolution of the eastern Mediterranean region. Bounded both in the north and the south by Tethyan suture zones, the BG remained as a fragments of a unique intrusive body surrounded by Tethyan seaways during the late Mesozoic. The BG (figure 1) is products of magmas evolved during the early stages of the collisional tectonics that affected much of Sakarya continent during the latest Mesozoic through early Cenozoic.

In this paper, we present the geology of the Late Cretaceous felsic plutons of Sakarya continent and new geochemical data from a series of BG, using a more functional classification scheme for the interpretation of regional tectonic settings and nature of magma.

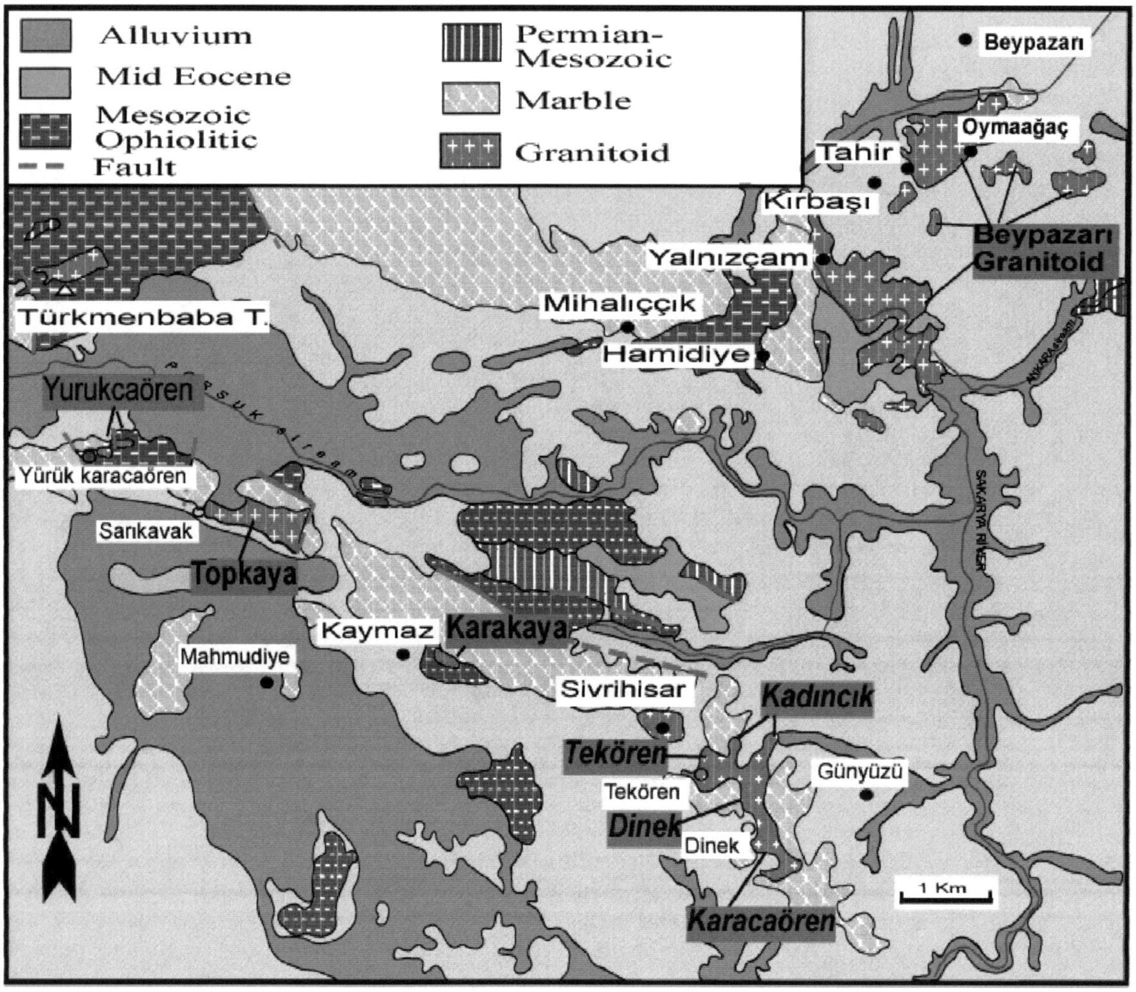

Figure 1. Regional geological map of Beypazari Granitoid (BG) in Sakarya continent.

2. Regional geology

The BG consists of the various felsic intrusive rocks out crops within the Sakarya continent intruded into the Eskisehir metamorphic rocks and Tethyan ophiolites ranging in age from the Late Cretaceous to the Miocene. Several sedimentary basins (Beypazari and Kirbasi) are initially evolved as peripheral foreland and/or forearc basins in the Miocene. The west and north part of the BG is occupied by the branch of Tethyan ophiolites. Protoliths of the high-grade rocks in the metamorphic massifs of the BG were Palaeozoic to Mesozoic platform assemblages. These sedimentary units underwent regional metamorphism as a result of burial beneath the Tethyan ophiolites derived from the Izmir-Ankara-Erzincan suture zone to the north and west [1]. Subsequent collisional events in the latest Mesozoic and early Cenozoic resulted in further crustal imbrications and thickening of crystalline rocks in the metamorphic of Eskisehir region.

The aeromagnetic anomaly map and field geology of BG reveal that the granitoid intruded into the Sakarya continent as the elliptical shaped body and surrounded by Kirmir stream at the North-North East and Ankara stream at South-South west (figure 2).

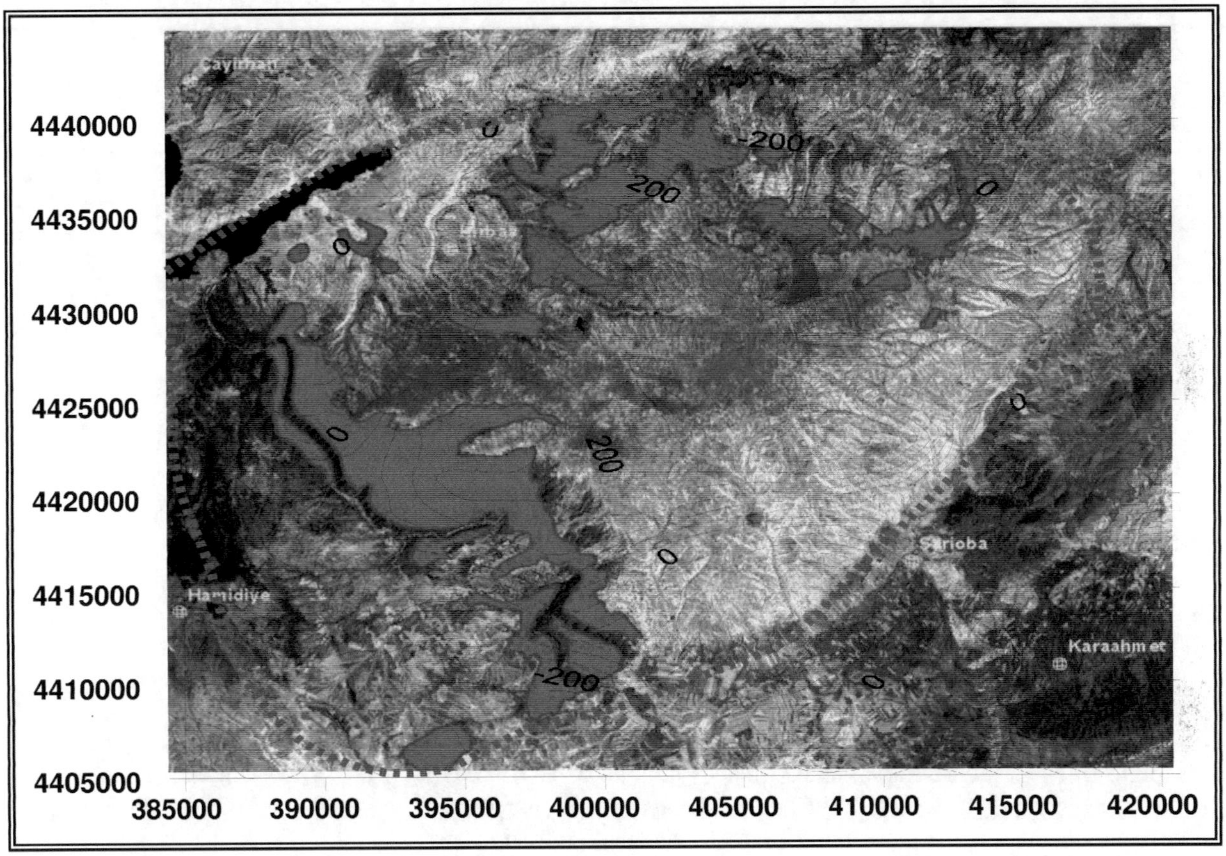

Figure 2. Aeromagnetic anomaly map indicating the boundary of BG on the landsat view. The dashed line shows the estimation boundary of BG, and the red area shows BG outcrops.

3. Geology and petrography of Beypazari granitoid

The BG intruded into the metamorphic and ophiolitic rocks of the Sakarya continent, consisting of felsic and mafic rocks produced by calc-alkaline in the latest Mesozoic (figure 1). The BG are composed mainly of is composed of five different units, these are monzonite, quartz monzonite, granodiorite, granite and alkali feldspar granite (figure 3). Granodiorite and granite are more abundant than monzonitic and are generally distributed at the inner zone of the exposures. Mafic plutonic rocks (mostly gabbros) are spatially associated and coeval with the felsic plutons with relatively small exposure bodies.

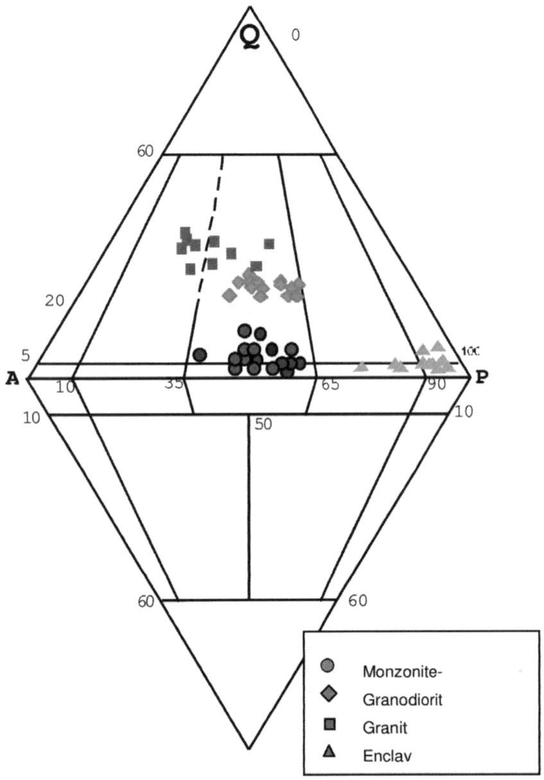

Figure 3. Nomenclature of the BG.

Granodiorite and granite occurs in the inner part of the region. They have a phaneritic texture and are light grey to pinkish grey colour in hand specimen. They have gradual contacts with monzonite and quartz monzonite. The lack of feldspar megacrysts in the granite and granodiorite make it possible to distinguish these units from the monzonite and quartz monzonite in the field. In the southern part of the BG, the monzonite has gradational contacts with the quartz monzonite. The presence of feldspar megacrysts and the abundance of amphibole are the main field criteria used to distinguish the monzonite from the quartz monzonite. All the subunits of BG is cut by dikes, a few cm to 3 m in thickness and composed of alkali feldspar granite and granite porphyry. The orientation of these dikes is commonly parallel to the general NW trend of the whole intrusive body [2].

The thin section the granite and granodiorite have a holocrystalline granular texture and is mainly composed of quartz, plagioclase, orthoclase, biotite with rare amount of amphibole. Titanite, epidote, zircon, apatite and opaque minerals are also observed as accessory minerals (figures 4a and b). Quartz forms the main felsic mineral and occurs as anhedral grains. Plagioclase is generally subhedral in shape, has an oligoclase composition, and displays albite twinning. Subhedral to anhedral biotites locally contain apatite and zircon inclusions and form large clusters.

The monzonite and quartz monzonite are exposed in the outer zone of the study area and has gradational contacts with the each other. The distinction between these two units is based largely on the quartz content and the texture governed by the high amount of mafic minerals. The contact of the quartz monzonite with the biotite granite body is also gradational, with the mafic mineral content decreasing from the former to the latter. In thin section, the monzonite and quartz monzonite have holocrystalline granular texture. Their mineral assemblage consists essentially of plagioclase, orthoclase, hornblende, and biotite with small amounts of quartz and pyroxene. The plagioclase in this unit has oligoclase and andesine compositions and is mostly subhedral and rarely euhedral in shape. Some hornblende crystals contain pyroxene relics in their core, whereas some appear to be locally replaced by biotite along fractures.

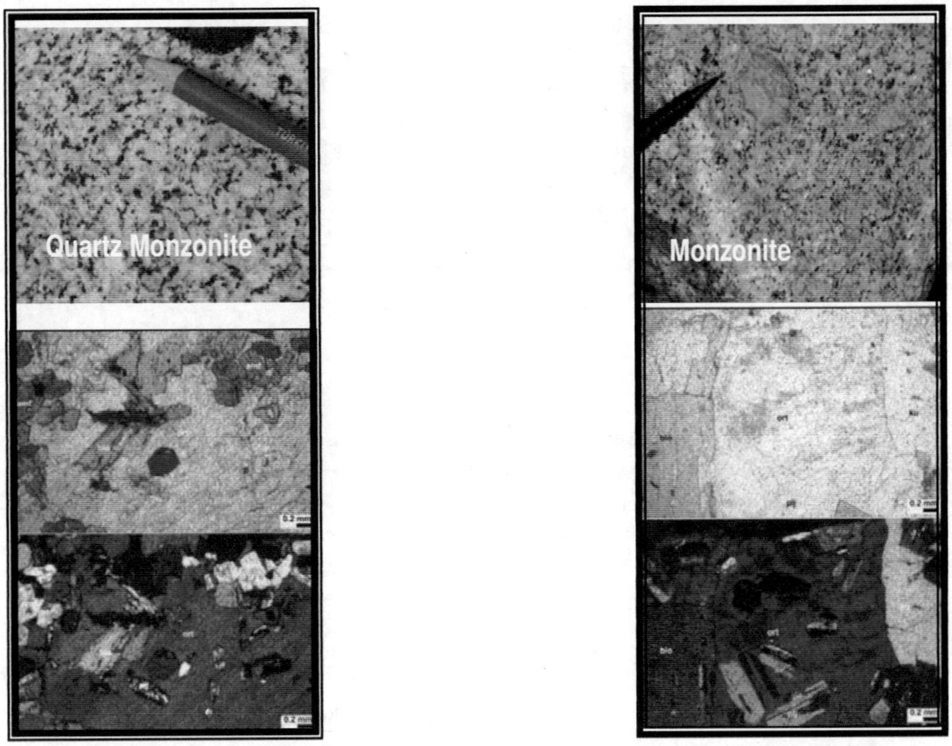

Figure 4. (A) Handspecimen and microphotograph of quartz monzonite (left) and monzonite (right) of BG.

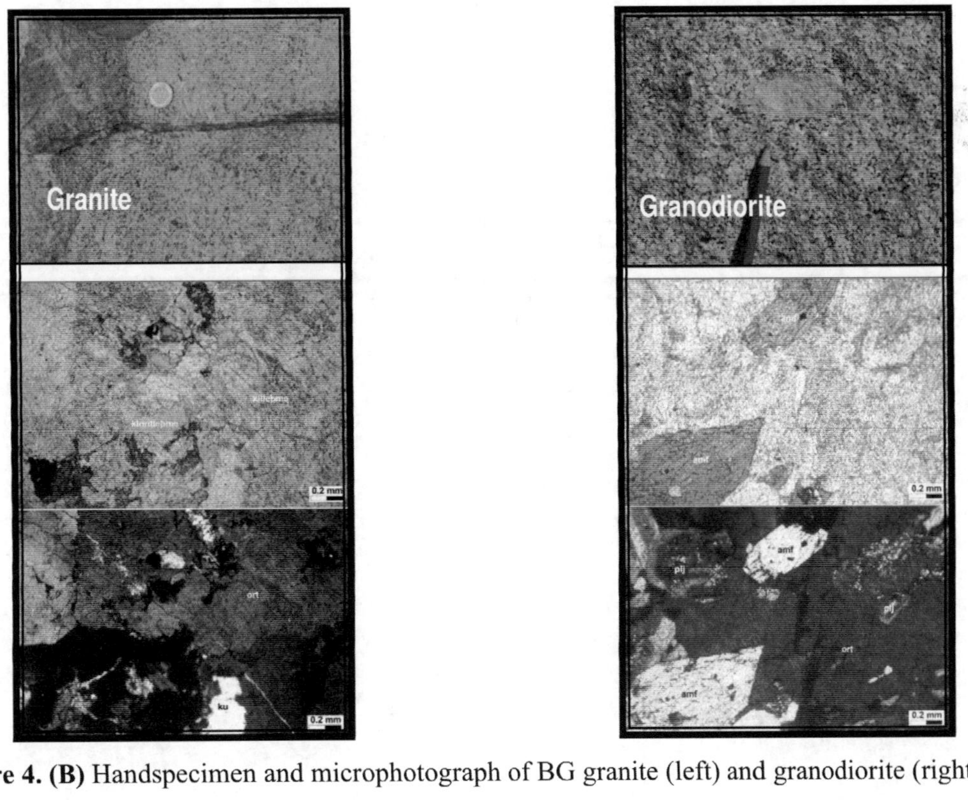

Figure 4. (B) Handspecimen and microphotograph of BG granite (left) and granodiorite (right).

4. Geochemistry of the Beypazari granitoid

Major and trace element analysis of a total of 63 representative samples were performed using PEDXRF and ICP in the petrography laboratory of Ankara University Geological Engineering department, Turkey. The correlation of BG with S, I, A, M and H type granites are given in Table 1. The results show that BG rocks are similar to the features of H type granite.

Table 1. Properties and comparison of Beypazari Granitoid (BG) with the other type of granitoids.

Granite type/Characteristic	S-type	I-type	M-type	A-type	H-type	BG
Al Saturation	Peralum.	Metalum.	Metalum.	Peralkaline	Metaalum. & Peralum.	Metalum.
Magma Character	Alk. & Calcalk.	Calcalk.	Calcalk. & Tholeiitic	Alk.	Calcalk.	Calcalk.
A/CNK	> 1.1	< 1.1	< 1.0	> 1.0	Similar	< 1.1
Variation diagram	Irr.	Reg.	Reg.	Reg.	Reg.	Reg.
$^{87}Sr/^{86}Sr_i$	> 0.707	< 0.705	< 0.705	Similar	Similar	Similar
$\delta^{18}O$	> 9 ‰	< 9 ‰	< 9 ‰	Similar	Similar	Similar
Mafic minerals	Mica	Amph. ± Mica	Amph. ± Mica	Amph. ± Mica ± Pyx	Amph Bt ± Pyx	Amph. Bt. ± Pyx
Titanite	±	Common	Common	Common	Common	Common
Apatite	Inclusion	Common	Common	Common	Common	Common & inclusion
Garnet	Mostly	-	-	± (Melanite)	-	-
Mafic Magmatic Enclave (MME)	-	+	+	+	+	+
References	[3]	[4]	[4] & [5]	[6]	[4]	This work

Notes: Abbreviations: BG, Beypazari Granitoid; Peralum., peraluminous; Metalum., metaluminous; Alk, Alkaline; Calcalk., Calcalkaline; Irr., irregular; Reg., regular; Similar, similar to other granites; Amph, amphibole; Pyx, pyroxene; "-", not present; "+", present; "±" varies depending on sample.

The SiO_2 contents of the analyzed rocks decrease systematically from the granite through the granodiorite to the monzonite, whereas the $Na_2O + K_2O$ contents increase in this order. On the total alkali versus silica (TAS) diagram, all the sub units of BG plot in the sub-alkaline field (figure 5a). Rocks of the granite to monzonite are calc-alkaline in nature (figure 5b) and display a linear trend suggesting a transitional change in composition. The mafic microgranular enclaves of these rocks have calcalkaline in compositions.

In order to see the possible genetic relations among the granite, granodiorite and monzonite samples, a major element variation diagram are plotted in Figure 6. There is a negative correlation on the plots of SiO_2 versus Fe_2O_3, TiO_2, MgO, CaO, and Al_2O_3, and a positive correlation on the plots of SiO_2 versus K_2O and Na_2O. This phenomenon may indicate that the evolution of the magmas of these three rock units involved crystal fractionation processes coupled with assimilation.

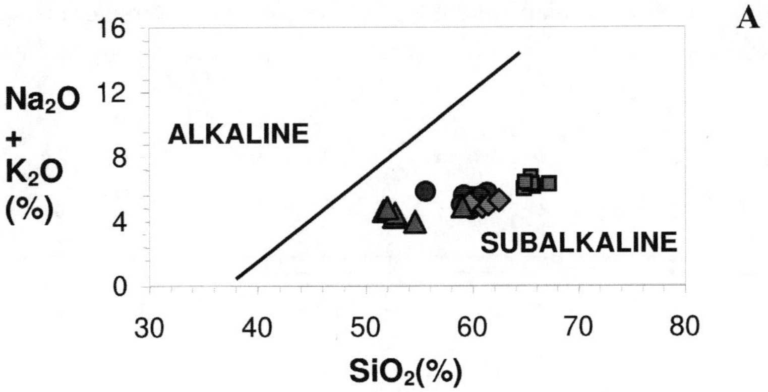

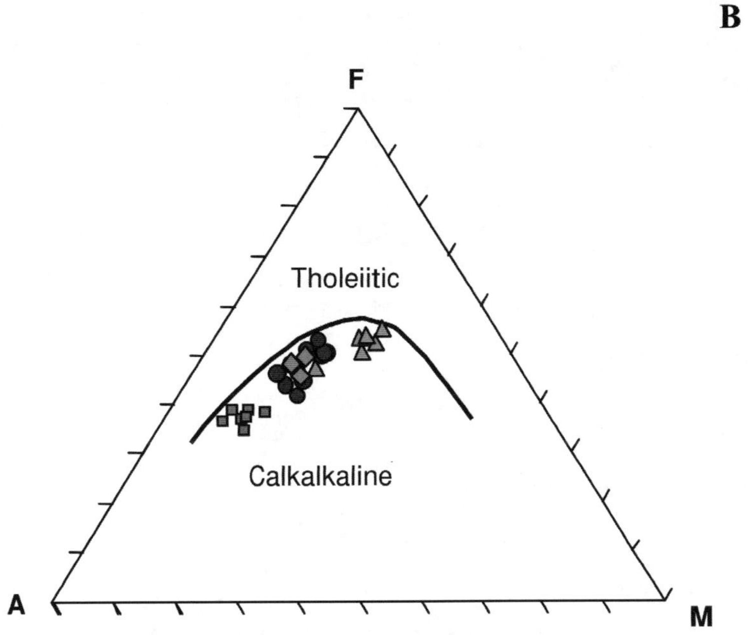

Figure 5. (A) Alkai versus SiO_2 and **(B)** AFM diagram of rocks from the BG.

Figure 7 shows the distribution of elemental patterns in rocks from each units on an ORG-normalized (hypothetical ocean ridge granite) diagram. In general, all subunits of the granite, granodiorite and monzonite show enrichment in large–ion lithophile elements (LILE) and depletion in high-field strength elements (HFSE) relative to ORG, similar to the patterns of rock units formed by subduction and/or collision tectonics [7].

The trace element discrimination diagrams of [8] (figures 8 and 9) and [9, 10]; are prepared in order to determine the tectonic environment in which the BG were emplaced. All subunits of the granite, granodiorite and monzonite plot in the VAG + COLG + ORG (volcanic arc granite + collision granite + ocean ridge granite). The Y versus Nd diagram is used to differentiate VAG + Syn-COLG, WPG and ORG affinities of the rocks, and the Y+Nb versus Rb diagram is used to differentiate between VAG and ORG, WPG and Syn-COLG affinities. We used the Rb versus Yb+Ta and Ta versus Yb diagrams to differentiate the VAG affinity of the analyzed rock samples from syn-COLG,

WPG, and ORG affinities. In these diagrams, subunits of the granite, granodiorite and monzonite plot mostly in the VAG field.

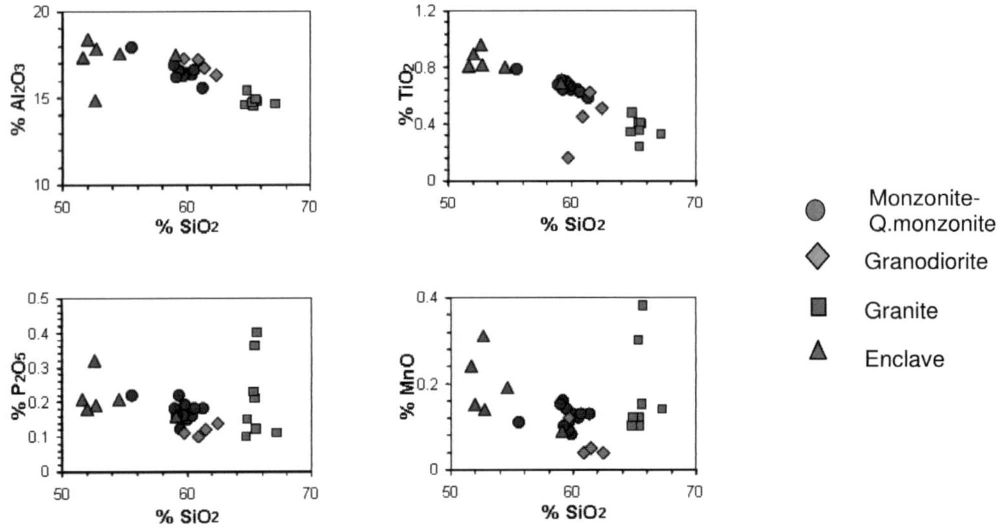

Figure 6. Variation diagrams of rocks from the BG.

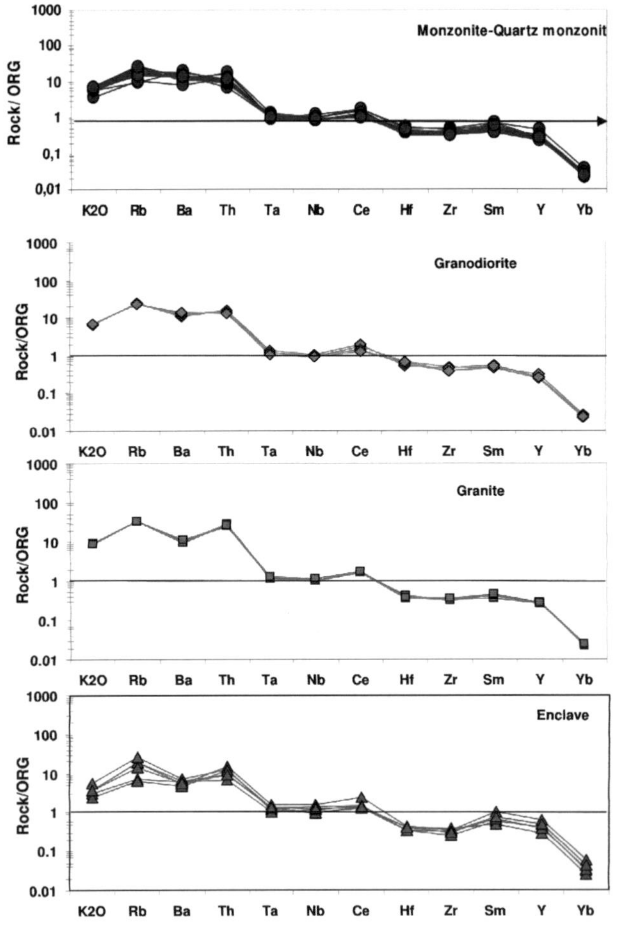

Figure 7. ORG normalized elemental patterns of rocks from the BG.

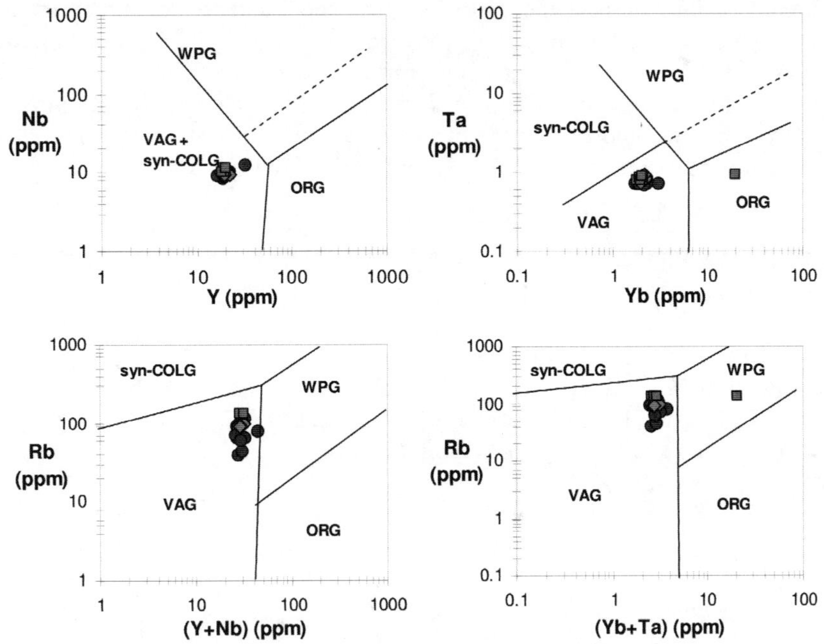

Figure 8. Tectonic discrimination diagrams of BG (fields from [8]).

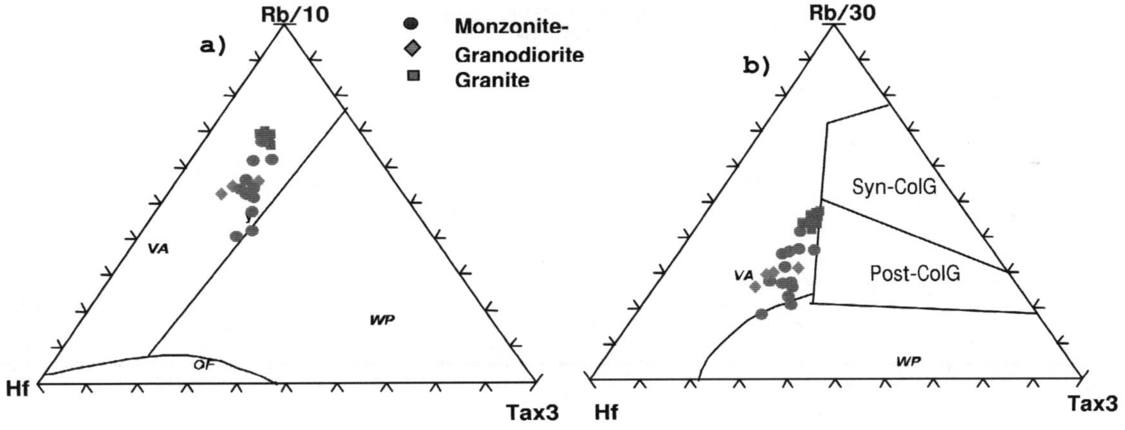

Figure 9. Tectonic discrimination diagrams (a) Rb/10, Hf, Ta x 3, (b) Rb/30, Hf, Ta x 3 of [9].

5. Conclusions

The Late Cretaceous rocks of BG were emplaced after the obduction of the Tethyan ophiolites, which were derived from the Ankara-Erzincan suture zone to the north and before its collisions with the Pontide and Tauride continental blocks during the Middle Eocene. Therefore, the magmatic evolution of these plutons preceded the terminal continental collisions in the region and hence should not be considered as post-collisional in nature.

The BG can be grouped into monzonite, quartz monzonite, granodiorite, granite and alkali feldspar granite based on their field occurrences and distinct differences in mineral and chemical compositions. All the subunits are calc-alkaline in nature ranging in composition from monzonite, quartz monzonite, granodiorite, granite and alkali feldspar granite. They show enrichment in LILE and depletion in HFSE relative to ORG. They display trace element signatures suggesting a crustal component derived from subduction-influenced source enrichment and crustal contamination resulted from fractional crystallization processes during magma transport through the Sakarya continent.

The field observations, mineralogy, petrography with the whole geochemical data reveal that the BG magmas are derived from a subduction-modified magma and metasomatized mantle source with considerable crustal contribution. The emplacement of the BG into the Sakarya Continent was a syn-collisional event during the terminal closure of the Inner-Tauride Ocean.

Acknowledgements
The authors gratefully acknowledged to AFP (03-K-120-190-4-1) and 2007-07-45056 of Ankara University for the supporting of the field work.

6. References

[1] Whitney D L and Dilek Y 1997 *Geology* **25** 1023–6
[2] Diker S, Çelik M and Kadıoğlu Y K 2006 *Environ Geol.* **11** 365–75
[3] Chappell B W and White A J R 1974 *Pac. Geol.* **8** 173–4
[4] Castro A 1991 *Earth Sci. Rev.* **31** 237–53
[5] Clarke D B 1991 *Granitoid rocks.* (New York: Chapman and Hall)
[6] Collins W J, Bearns S D, White A J R and Chappell B W 1982 *Contrib. Mineral. Petrol.* **80** 189–200
[7] Hezel D C 2006 *Geochim. Cosmochim. Acta* **70** 1548–64
[8] Pearce J A, Harris B W and Tindle A G 1984 *J. Petrol.* **25** 956–83
[9] Harris P G, Kennedy W Q and Scarfe C M 1986 Volcanism versus plutonism-the effect of chemical composition. Liverpool Geological Society Symposium: Mechanisms of Igneous Intrusion
[10] Hopep G A, Woodsy R and Munce C G 2001 *Min. Eng.* **14** 1565-77

IOP Conf. Series: Earth and Environmental Science **2** (2008) 012015 doi:10.1088/1755-1307/2/1/012015

Trace element and isotopic composition of mafic and ultramafic cumulate xenoliths in alkaline basalts from the Eastern Rhodopes, Bulgaria: Inferences on deep processes under the metamorphic core complexes

P Marchev[1,4], S Arai[2], Y Ishida[2], M Shirasaka[2] and H Downes[3]
[1]Department of Geochemistry and Petrology, Geological Institute of Bulgarian Academy of Sciences, 1113 Sofia, Bulgaria
[2]Department of Earth Sciences, Graduate School of Natural Sciences and Technology, Kanazawa University, Kakuma, Kanazawa 920-1192, Japan
[3]School of Earth Sciences, Birkbeck, University of London, Malet Str., London WC1E 7HX, UK

E-mail: pmarchev@geology.bas.bg

Abstract. Oligocene within-plate alkaline basalts, intruded in the Eastern Rhodope metamorphic core complexes Kesebir and Biala Reka, carry ultramafic and mafic xenoliths, consisting of clinopyroxenites, websterites, orthopyroxenites, 2-pyroxene- and clinopyroxene gabbros. These xenoliths are interpreted as samples of layered intrusions formed at upper mantle-lower crustal depths. Calculated parental liquids from LA ICP-MS analyses of clinopyroxene in the ultramafic clinopyroxenites, orthopyroxenite and websterite indicate that they may have formed from melts similar in composition to their host basanitic magma. The clinopyroxene REE patterns of least magnesian websterites suggest amphibole and clinopyroxene fractionation. Trace element compositions of gabbros and high-Fe clinopyroxenites indicate that they are the most differentiated product from the same parent. Calculated parental liquids for some of them, however, show geochemical features typical for subduction-related rocks. In addition, Sr and Nd isotopic compositions of evolved websterites and gabbros are more radiogenic than those of the host basanites. This transition from within-plate to subduction-related signature can be related to fractionation of amphibole, accompanied by crustal contamination of the basanitic magma with more radiogenic lower crustal material.

1. Introduction

Ultramafic to mafic xenoliths are described from within-plate alkaline basalts erupted through the metamorphic core complexes of the SE California and W Arizona [1]), the Menderes Massif [2, 3] and the Eastern Rhodope Massif [4]. Those three areas exhibit many structural and magmatic evolutionary similarities, discussed in [4, 5]. The authors concluded that the ultramafic and mafic xenoliths might be the products of mafic magma underplating, which caused thermal modification and extension of the lower and middle crust in the metamorphic core complexes. Thus, these rocks are an attractive potential source for the heat required to prove the thermal and experimental models [6, and references

[4] To whom any correspondence should be addressed.

IOP Conf. Series: Earth and Environmental Science **2** (2008) 012015 doi:10.1088/1755-1307/2/1/012015

therein] that predict that underplating of hot magma at the base of the crust may significantly modify its thermal and mechanical properties, enhancing deformation and strain localization in an extensional environment.

Marchev et al. [4] provided preliminary major element, mineral composition and textural data on the largest occurrence of ultramafic-mafic xenoliths, the Purjenaka dyke, as well as P-T estimates of the equilibration condition of their crystallization. Thermobarometric studies and comparison with experimental works suggest that the cumulate sequence formed from mafic magma at pressures of 14-9 kb (45-30 km) and temperatures of 1200-850 °C. However, the composition of the primary magmas and the evolutionary processes of the rocks have remained largely unknown.

In this work, we present extended LA ICP-MS trace element analyses of minerals from ultramafic and mafic xenoliths, along with limited whole rock ICP and Sr and Nd isotope analyses, in an attempt to understand the nature of their sources, the genetic relationships between the different cumulitic rocks and the role of crustal contamination in their genesis. We show that the most primitive cumulitic rocks are segregations from crystallizing alkaline melts. More evolved websterites and gabbroic rocks suffered extensive transformation through fractionation and assimilation of crustal material.

2. Geological background

The Eastern Rhodopes occupy the eastern part of the Rhodope Massif, a complicated structure formed during the Variscan and Alpine orogenesis. The regional-scale tectonic setting is dominated by two late Alpine metamorphic core complexes, namely the Kesebir-Kardamos and the Biala Reka-Kechros domes [7, 8, 9] (figure 1). Development of the core complexes started at the Late Cretaceous. Extension along continuously mappable low-angle detachment faults exposed the lower high-grade

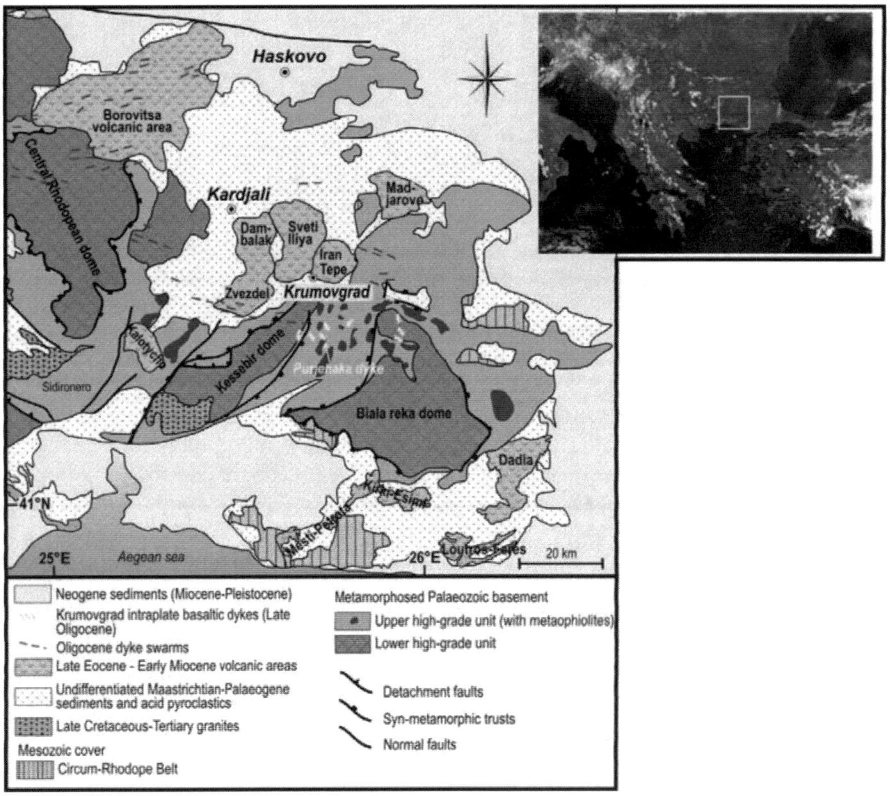

Figure 1. Schematic geologic map of the Eastern Rhodopes taken from [4]. Inset shows position of the Rhodope Massif in SE Europe.

unit in the core, structurally flanked by the lithologies of the higher high-grade unit and the overlying low-grade Mesozoic unit. The latest period of extension was expressed by high-angle faulting, sedimentary basin formation and widespread Late Eocene-Oligocene volcanism (39?-26 Ma). The metamorphic rocks of the core complexes themselves were intruded by numerous dykes, which can be divided into two groups [5]. The older one (~32 Ma) is represented by bimodal acid (rhyolite-rhyodacite) and intermediate (basaltic-andesite-latite) compositions. The younger dykes in the Kessebir and Biala Reka domes are of Middle Oligocene age (28-26 Ma). They are of intraplate affinity and have been referred to as the Krumovgrad alkaline basalts (KAB) [10].

3. Krumovgrad alkaline basalts (KAB) and xenoliths

The dykes are located 20-30 km SE of the town Krumovgrad (figure 1), cropping out over an E-W-oriented area of 1000 km^2 in the Biala Reka and Kessebir domes. Chemical and isotopic compositions of the dykes were reported in [4, 10]. Dykes are represented by basanites, alkali basalts and lamprophyres. The rocks have high $^{143}Nd/^{144}Nd$ (0.51290-0.51289), low $^{87}Sr/^{86}Sr$ (0.70323-0.70338) and high $^{206}Pb/^{204}Pb$ (19.02-18.91) at relatively low $^{207}Pb/^{204}Pb$ (15.64-15.52) ratios, interpreted to indicate a source similar to the European asthenospheric reservoir [11].

Almost all of the xenoliths come from the largest dyke of KAB, outcropping SE of Egrek village, named Purjenaka. The xenoliths are of three dominant types: (1) mantle peridotites; (2) ultramafic and mafic cumulates; (3) metamorphic rocks from the Rhodopean metamorphic basement, with predominance of the types 2 and 3 [4].

4. Petrography and analythical methods

Minerals from 17 cumulitic samples, from those studied in [4], have been analysed in this paper for trace elements. On the basis of clinopyroxene composition and associated minerals, the following subdivision of the rocks has been made: augite clinopyroxenites (IIEG7; IIEG18; Bz24-3c; Eg06-3b); salite clinopyroxenites (Bz24); ore (ilmenite-sulfide) clinopyroxenites (Eg97-4a), and hornblende-pyroxenites (Bz24-5a; 26-2a). Sample Bz24-3c contains small amount of pseudomorphs after olivine, showing transition to wehrlite. Websterites can be divided into olivine websterites (Bz24-3), websterites (EG97-4) and amphibole websterites (Eg97-1 and IIEG99-2). Gabbros are represented by augite gabbro (Bz03-2/5 and Bz24-5c) and 2-pyroxene gabbro (Bz-24-5b). A composite xenolith (Bz03-2/6), consisting of gabbro and clinopyroxenite layers, is the only studied layered xenolith.

Trace-element abundances in the major phases of the cumulitic rocks (olivine, orthopyroxene, clinopyroxene, amphibole, plagioclase and apatite) were determined in the University of Kanazawa, Japan; ETH Zurich, Switzerland; and the University of Perugia, Italy. The analyses in Kanazawa were made using a LA system (GeoLas Q-Plus, MicroLas) coupled to an ICP-MS system (Agilent 7500s, Yokogawa Analytical Systems). LA-ICP-MS analyses of minerals in ETH, Zurich were determined using an ArF excimer UV (193 nm) laser source coupled with an ELAN 6100 ICP quadrupole mass spectrometer. The analyses at the University of Perugia were performed using a Thermo Electron X7 quadrupole based ICP-MS coupled to a New Wave UP213 frequency quintupled Nd-YAG laser.

Trace-element compositions of the 2 large xenoliths (a websterite and a gabbro), earlier analysed for major element analyses [4], were analyzed by ICP-MS, X7, Thermo Electron at the Department of Earth Sciences at the University of Kanazawa. The chemical composition of another gabbro sample is calculated by mass balance based upon the modal composition and mineral major element. Sr and Nd isotope composition of one of the large xenoliths was analysed at the Royal Holloway University of London with a VG354 TIMS. The second one was analysed at the University of Florence with a Finnigan MAT 262 TRITON. Sr and Nd isotopic results for the KAB are those reported in [10].

5. Trace element chemistry of minerals

On the basis of their chondrite-normalised (C-normalised) profiles, we can roughly identify 4 groups of clinopyroxenes. Group I includes clinopyroxenes of clinopyroxenites IIEg18 and Eg06-3b, amphibole clinopyroxenite 24-5a and 26-2a. REE profiles of these clinopyroxenes display convex-upward shape peaking at Nd and Sm (figure 2a). The shape of the profiles is very similar to those of the phenocrysts core and Ti-augite rim of host basanites (e.g. IIEg-7), with the latter being considerably more enriched. REE patterns for clinopyroxene of orthopyroxenite 24-1, olivine websterite 24-3; websterite Eg97-4; clinopyroxenite 24-3c, IIEG7, and 24-5c (group II) also exhibit similar convex-upward REE patterns (figure 2b). Unlike the previous group, however, they have peaks at Ce with some of them showing a small Eu* anomaly (e.g. 24-1, 24-5c; Eg97-4). The primitive mantle-normalized (PM-normalised) trace element patterns of these two groups are comparatively similar, with troughs in Nb, Pb, Sr, Zr; Ti has troughs only in the more Fe-rich augites, but is missing in the most Mg-rich clinopyroxenes. Group III includes the pyroxenes of most Fe-rich websterites. They show overall depletion in trace-elements and REE patterns having mildly concave-upward to "W"-shaped REE profile, with one sample with a positive Eu* anomaly (figure 2c). PM-normalised trace element patterns are close to 1 with small troughs in Nb, Zr, Hf, Ti. Group IV includes the most Fe-rich salitic and augite clinopyroxenes (figure 2d). They can be subdivided in 2 groups. Samples from layered pyroxene gabbro and clinopyroxenite display convex-upward REE patterns similar to the other clinopyroxenes, but with a pronounced Eu* anomaly, whereas clinopyroxenes of ore-clinopyroxenite and 2-pyroxene gabbro show highly enriched convex-upward MREE and very strong Eu* anomaly. The most unusual REE patterns are displayed by the core of clinopyroxene gabbro Bz03-2/5 which shows steep enrichment from La to Eu and nearly flat patterns to Lu. Similarly to the previous groups, they show troughs in Nb, Zr, Hf, Ti on their PM-normalised profiles.

Amphibole in the websterite Eg97-1 and clinopyroxenite IIEg7 has particularly high concentrations of Cr (4300 and 2300 ppm) and Ni (287-355 ppm), much higher than those in the coexisting clinopyroxenes. All amphiboles in the hornblende clinopyroxenites exhibit convex-upward REE shape, with peaks at Nd, similar to those of clinopyroxenites of group I and amphibole megacrysts (figure 3a). Similarly to the clinopyroxenes, high-Fe websterites have a "W"-shape profile, suggesting growth from strongly differentiated melt. PM-normalized trace element patterns of these amphiboles show high concentrations of incompatible trace element with peaks at Ba, Nb, Ta, Sr and Ti, a deep trough at Pb and a slight trough at Hf and Zr.

Orthopyroxene has significant concentrations of Ni, Co, V. Its C-normalized REE patterns (figure 3b) exhibit a continuous depletion trend from HREE to LREE, with variations up to an order of magnitude. These variations are similar to those of clinopyroxene and amphibole and obviously reflect fractional crystallization. Some of them show slight negative Eu* anomaly. The orthopyroxene of 2-pyroxene gabbro, which has the highest REE contents, shows the strongest negative Eu* anomaly, reflecting fractionation of plagioclase. PM-normalised profiles of the orthopyroxenes exhibit peaks in Th, Pb and troughs in Ba, Sr, and Sr.

Plagioclase is a major phase in the gabbroic rocks. It shows high concentrations of LREE, Sr, Ba and Pb REE patterns. Its C-normalized REE patterns (figure 3c) are characterised by a steep slope from La to Sm and strong positive Eu* anomaly. The PM-normalized trace element patterns for the plagioclases are characterised by a significant enrichment in Rb, Ba, La, Eu, Pb and Sr and negative Nb, Zr, Hf and Ti anomalies. Anorthoclase outer zone of the plagioclase in the 2-pyroxene gabbro has high contents of Rb, Sr and, particularly, Ba.

The cumulitic apatite has strongly enriched REE contents, with C-normalised concentrations falling steadily from La to Lu (figure 3d) and pronounced Eu anomaly, which mimics that in the coexisting clinopyroxenes. Apatite included in amphibole megacrysts shows lower REE contents and no Eu anomaly. On the PM-normalized diagram, cumulitic apatite shows depletions in Sr, Nb, Pb Zr and Ti. Apatite enclosed in pargasite phenocrysts has overall lower contents of trace elements, suggesting a strongly fractionated nature of the clinopyroxenite rocks.

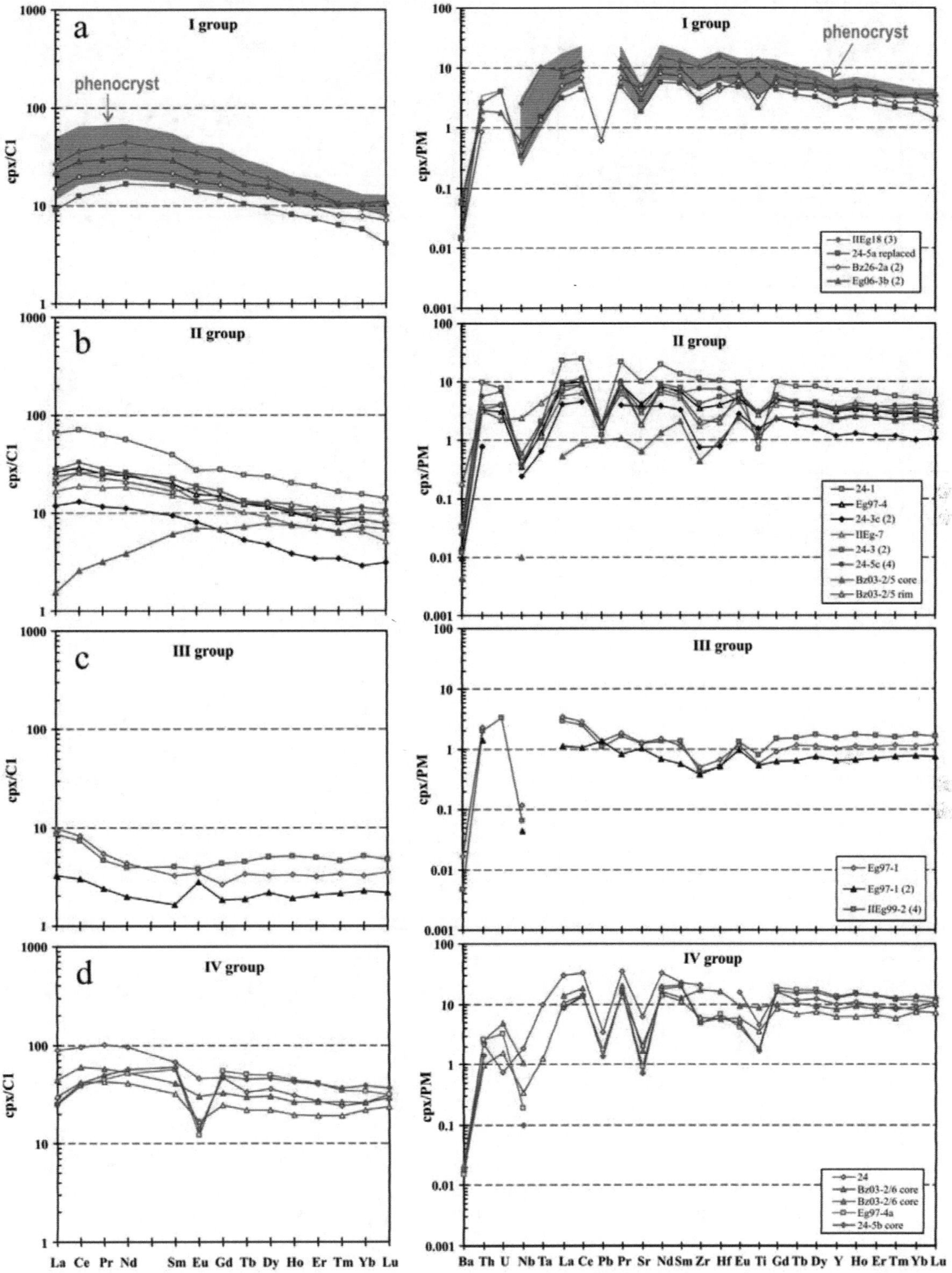

Figure 2. Chondrite-normalised and Primitive mantle-normalised trace element diagrams of clinopyroxenes. Shaded area – host basalt phenocryst.

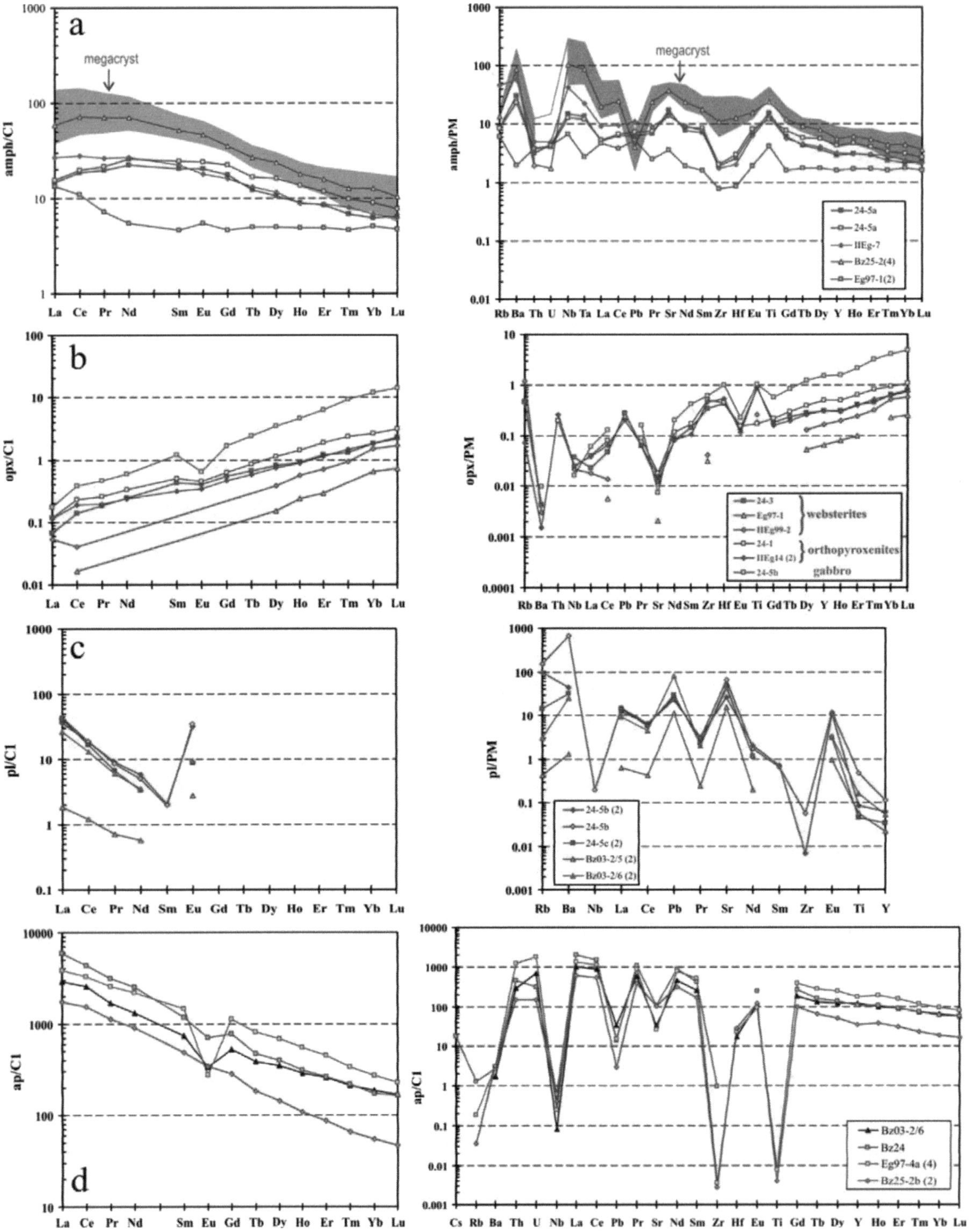

Figure 3. Chondrite-normalised and Primitive mantle-normalised trace element diagrams of amphibole, orthopyroxene, plagioclase and apatite. Shaded area – host basalts amphibole megacryst.

6. Whole rock chemistry of websterite and gabbro

Trace-element compositions for websterite IIEG99-2 and gabbro Bz03-2/5, and the calculated composition of Bz24-5c are shown in figure 4. REE abundances are low, with C-normalized REE patterns characterized by linear MREE and HREE patterns and mild LREE enrichment. Gabbro samples shows slight to significant (Bz03-2/5) positive Eu anomaly, reflecting plagioclase accumulation. PM-normalized patterns show slight enrichment in LILE (Rb, Ba, Th, U) and the LREE (except Bz03-2/5), relative to the HFSE (Nb, Ta, Ti, Zr, Hf) and HREE. Gabbros exhibit also large Pb and Sr positive anomalies, weaker in the websterite. The shape of these patterns, with negative Nb, Ta and Ti anomalies and a positive Pb anomaly, is similar to subduction-related magmas.

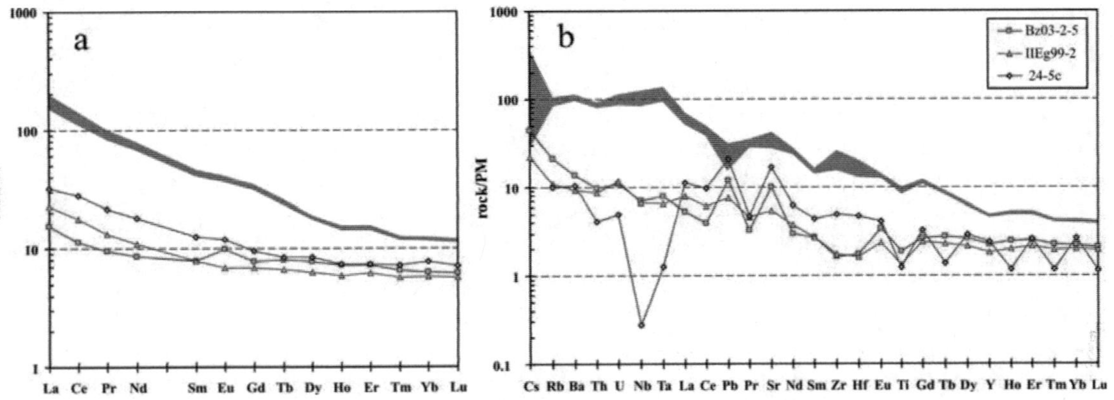

Figure 4. Chondrite-normalized (a) and primitive mantle-normalised (b) patterns of websterites and gabbros cumulites, compared with their host alkaline basalts (shaded area).

7. Isotope compositions

Sr and Nd isotope compositions of websterite Eg99-2 and gabbro Bz03-2/5 are reported in figure 5. Initial isotopic ratios were calculated for an age of 28 Ma. The websterite has Sr and Nd isotope ratios 0.70436 and 0.512820, respectively, whereas the gabbro shows slightly higher Sr and Nd isotope ratios (0.70444; 0.512879, respectively). These values are considerably more Sr radiogenic than those of the host basanites (0.70323-0.70338) but only slightly less radiogenic in Nd (0.512894-0.512902).

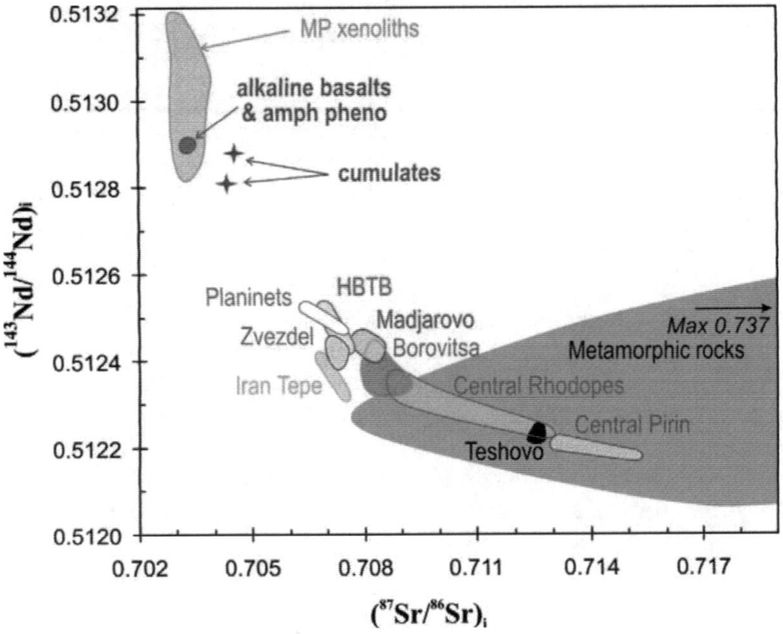

Figure 5. Sr vs. Nd isotope composition of cumulitic rocks compared to the mantle peridotites, alkaline basalts and amphibole phenocrysts, and Eocene-Oligocene orogenic igneous rocks and metamorphic basement.

8. Discussion

8.1. Origin of clinopyroxenites

All ultramafic clinopyroxenites, except high-Fe varieties, exhibit convex-upwards C-normalized REE patterns similar to those of clinopyroxene mega/phenocrysts and their Ti-augite rims. Such chondrite-normalised patterns are typical of clinopyroxene megacrysts crystallized from basic melts at high pressure [12]. Therefore, these characteristic suggest a magmatic origin of pyroxenes as cumulates from basic alkaline melts. In order to check this possibility, we have calculated the trace-element composition of the hypothetical melts from which they have crystallized using cpx/melt partition coefficients reported in [13] and [14]. Chondrite-normalised and primitive mantle-normalized diagrams for trace-elements of the calculated melts closely match the patterns of basanites (figure 6). The results strongly support the suggestion that the clinopyroxenites originated from melts similar in composition to the host Krumovgrad alkaline dykes.

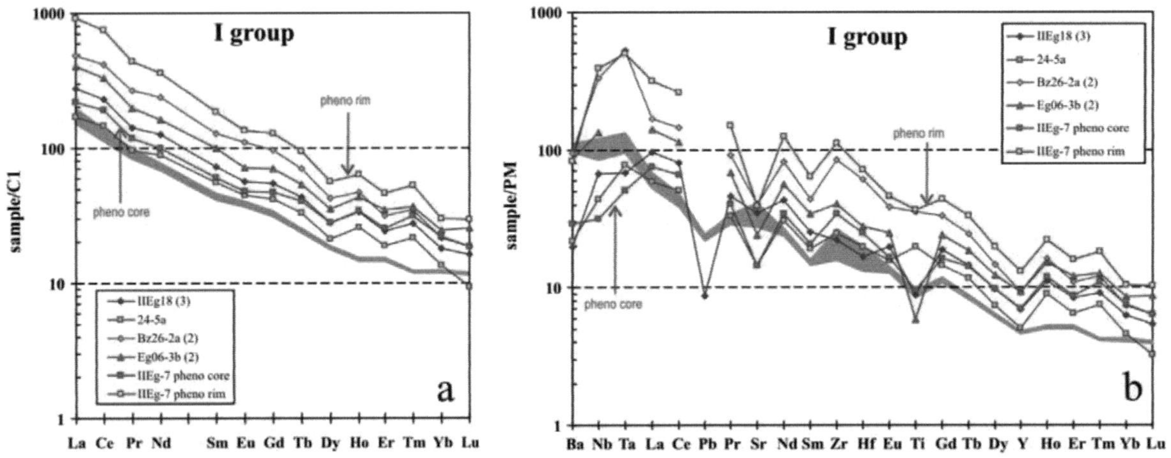

Figure 6. (a) Calculated chondrite-normalized REE patterns and **(b)** Primitive mantle-normalized trace element patterns of parental magma in equilibrium with clinopyroxenes from clinopyroxenites group I, using the partition coefficients of [12 and 13], compared with host alkaline basalts (shaded area).

8.2. Origin of orthopyroxenites and websterites

The origin of orthopyroxenites and orthopyroxene-rich websterites is controversial. The absence of composite xenoliths and relict grains of olivine and spinels in the orthopyroxenite, together with scarce presence of peridotitic xenoliths within the Eastern Rhodope xenoliths, suggest that the most Mg-rich orthopyroxenite Bz24-1 (Opx and Cpx Mg# ~89) is not the result of reaction with peridotites. The clinopyroxene REE profiles of the orthopyroxenite Bz24-1 and olivine websterite (Bz24-3) suggests that they might have crystallized from an alkali-basaltic melt, but the slight Eu anomalies suggest either plagioclase fractionation or contamination by a crustal component. Melts from which the orthopyroxene-rich rocks formed are believed to have Mg-Si-rich composition. Such magmas could be either boninite-like silicate melts strongly enriched in Sr and Pb [15, 16, 17], calc-alkaline magmas [18], or strongly crustal contaminated mantle-derived magmas [19]. The clinopyroxenes from the least Mg- websterites IIEg99-2 and Eg97-1 with Mg# 76-77 have flat to "W"-shape REE profiles (figure 2c), very similar to those of the amphibole crystals of Eg97-1 enclosed in the clinopyroxene (figure 3a). According to [20], such profiles are the result of strong differentiation, producing depletion in the MREE, flat negative slope and slight positive Eu anomaly.

8.3. Formation of ore clinopyroxenite and gabbros

These iron-rich rocks are most probably the results of early crystallization of high-Mg phases, which led to the formation of more Fe- and Al-rich magmas. Most of them achieved apatite saturation, which comprise considerable volumes in Bz03-2/5 and Bz03-2/6. Clinopyroxene and apatite trace-element compositions of the ore clinopyroxenite and 2-pyroxene gabbro, having deep Eu anomalies on their C-normalized profiles, are consistent with growth after a significant amount of plagioclase crystallization and depletion of the liquid in Sr and Eu.

9. Crustal contamination

Figure 6 presents Sr-Nd isotope ratios of the two analysed cumulitic samples, compared with the host lavas, orogenic lavas and metamorphic basement in the Rhodope Massif and mantle xenoliths from north Bulgaria. The cumulitic rocks have slightly higher Sr and almost identical Nd isotopes than their host rocks, but their isotopic characteristics differ significantly from those of the orogenic rocks and the Rhodopean metamorphic basement. It is obvious that their isotopic composition cannot be explained by mixing of a depleted asthenospheric component with the local mantle lithospheric component. The most reasonable explanation for the observed offset of the isotopic compositions of the cumulates from the alkaline basaltic composition, is interaction with lower/middle crustal wall-rocks. The composition of the lower crust in the area is not known, but if we assume an assimilant similar to the upper-middle crust in the area, with (Sr) = 300 ppm and $^{87}Sr/^{86}Sr$= 0.7200 [21], addition of ~8-9 wt% bulk crust is sufficient to explain the isotopic offset between the host basanites and cumulitic rocks. This estimate was calculated with assumption that primary magma had 750 ppm Sr. Such degree of contamination will have a comparatively small impact on the major and trace element budget of the cumulitic magmas.

10. Transition from intraplate to subduction-related signature

Depletion of Ta and Nb relative to other incompatible elements in the orogenic magmas has been ascribed to many processes: (1) Crystal fractionation of Fe-Ti oxides in the crust; (2) Presence of Ti-rich phases, e.g., rutile or sphene in the mantle wedge; (3) Relative immobility of Ta and Nb compared to REE and other elements in aqueous fluids derived from subducted material; (4) Inherited low Ta and Nb from subducted sediments; (5) Fractionation of Ti-rich hydrous silicates such as phlogopite or amphibole in the mantle or crust.

Petrography of the Eastern Rhodope cumulates [4] shows that amphibole is widely distributed in the mafic websterites and clinopyroxenite, either as an early phase or late intercumulus phase. PM-normalised patterns of the amphibole (figure 3a) are characterized by strong enrichment of Ti and Nb and Ta. Thus, early and progressive fractionation of clinopyroxene, orthopyroxene ± olivine and amphibole from a basanite parent melt may generate derivative gabbroic magmas depleted in Ti, Nb and Ta.

11. Conclusions

1) Trace element studies of most primitive cumulitic xenoliths from the Krumovgrad alkaline basalts are consistent with their formation from mantle-derived parent melts similar to their host alkaline magma.
2) The progressive change in the mineral composition in more evolved lithologies is consistent with extensive transformation through fractionation.
3) The preferential partitioning of Nb-Ta and Ti into amphiboles may deplete the most fractionated melts, producing subduction-related signatures.
4) The small shift of the isotopic composition of the cumulitic rocks from the alkaline basalts is most probably the result of a small percentage of assimilation of lower crustal material.

Acknowledgements

This work was supported by a grant of Japan Society for Promotion of Science (JSPS) to PM.

12. References

[1] Wilshire H G 1990 *J. Geophys. Res.* **95** 649–65
[2] Çakir U, Aydar E, Sen E and Bayhan H 1999 *J. Conf. Abstr.* **4**
[3] Holness M B and Bunbury J M 2006 *J. Volcanol. Geotherm. Res.* **153** 241–61
[4] Marchev P, Arai and Vaselli O 2006 *Geol. Soc. Am. Spec. Paper* **409** 237–58
[5] Marchev P, Raicheva R, Downes H, Vaselli O, Chiaradia M and Moritz R 2004 *Tectonophysics* **393** 301–28
[6] Corti G, Bonini M, Innocenti F, Manetti P, Conticelli S, Innocenti F, Manetti P and Sokutis D 2003 *Earth. Sci. Rev.* **63** 169–247
[7] Burg J-P, Ricou L-E, Ivanov Z, Godfriaux I, Dimov D and Klain L 1996 *Terra Nova* **8** 6–15.
[8] Krohe A and Mposkos E 2002 *Geol. Soc. London Spec. Pub.* **204** 151–178.
[9] Bonev N 2006 *Geol. Soc. Am. Spec. Paper* **409** 211–35
[10] Marchev P, Vaselli O, Downes H, Pinarelli L, Ingram G, Rogers G and Raicheva R 1998 *Acta Vulcanol.* **10** 233–42.
[11] Wilson M and Downes H 1991 *J. Petrol.* **32** 811–49
[12] Irving A J and Frey F A 1984 *Geochim. Cosmochim. Acta* **48** 1201–21.
[13] Hart S R and Dunn T 1993 *Contrib. Mineral. Petrol.* **113** 1–8
[14] Wood B J and Blundy J D 1997 *Contrib. Mineral. Petrol.* **129** 166–81
[15] Bali E, Falus G, Szabó C, Peate D W, Hidas K, Torok K and Ntaflos T 2007 *Mineral. Petrol.* **90** 51–72
[16] Santos J F, Schärer U, Gil Ibarguchi J I and Girardeau 2002 *J. Petrol.* **43** 17–43
[17] Cvetković V, Lazarov M, Downes H and Prelević D 2007 *Lithos* **94** 90–110
[18] Féménias O, Coussaert N, Bingen B, Whitehouse M , Mercier J-C and Demaiffe D 2003 *Chem. Geol.* **199** 293–315
[19] Wang CY, Zhou M-F and Keays R R 2006 *Contrib. Mineral. Petrol.* **152** 309–321
[20] Meurer W P and Claeson D T 2002 *J. Petrol.* **43** 607–29
[21] Peytcheva I 1997 *Rev. Bulg. Geol. Soc.* **58** 157–65 (in Bulgarian with English abstract)

Structural and $^{40}Ar/^{39}Ar$ age constraints on the Kulidjik nappe: A record of an early Alpine thrust tectonics in the northeastern Rhodope Massif, Bulgaria

Nikolay Bonev[1,4], Richard Spikings[2], Robert Moritz[2] and Peter Marchev[3]
[1] Sofia University "St. Kliment Ohridski", Department of Geology and Paleontology, Sofia 1504, Bulgaria
[2] Department of Mineralogy, University of Geneva, CH-1205 Geneva, Switzerland
[3] Department of Geochemistry and Petrology, Geological Institute, Sofia 1113, Bulgaria

E-mail: niki@gea.uni-sofia.bg

Abstract. In the northeastern Rhodope Massif, a unique juxtaposition of assumed high-grade basement allochthon onto a low-grade Mesozoic unit is exposed in the Kulidjik nappe. The thrusting is interpreted as post-Jurassic to pre-Late Eocene. The tectonostratigraphy comprises the following units: (i) an upper unit of high-grade basement of the eastern Rhodope consisting of various amphibolite facies rocks with metaophiolite lenses; (ii) an overlaying Mesozoic (including Jurassic radiolarians) low-grade unit; (iii) a structurally overlaying allochthon of the Kulidjik nappe; and (iv) unconformable Late Eocene sedimentary and Oligocene volcanic cover rocks. The basement unit contains a NNE-SSW oriented lineation rarely associated with top-to-the NNE ductile shearing. Greenschist facies retrogression of this unit is diagnostic of local thrusting of the overlying low-grade unit. The greenschists of the low-grade unit represent arc metavolcanic rocks overlain by phyllites. The structural elements in this unit have similar attitudes as the basement unit. Kinematic indicators in the low-grade unit demonstrate internal top-to-the NNE ductile shearing during greenschist facies conditions. The allochthon consists of klippen composed of muscovite-albite gneisses. The flat-lying thrust contact with ductile-brittle deformation contains mylonites and cataclasites. Previously reported local top-to-the SSW ductile shearing in the allochthon is probably a consequence of shear instabilities. Deformation-crystallization relationships revealed fabrics in the allochthon compatible with greenschist up to lower amphibolite facies conditions without retrogression and similar ENE-verging fold patterns in the low-grade unit and the allochthon at the thrust contact. Two muscovite samples from distinct klippen yielded inverse isochron $^{40}Ar/^{39}Ar$ ages of 156.07±1.30 Ma and 162.62±2.69 Ma, which reflect cooling below 350°C of the allochthon associated with its exhumation during emplacement after greenschist-lower amphibolite facies metamorphism. Structural data indicate NNE-directed tectonic transport of the low-grade footwall unit, which is consistent with emplacement in comparable metamorphic conditions, and the thrust direction of the allochthon in the hanging wall. The subduction-accretionary origin of the low-grade unit and the new age results allow us to link this tectono-metamorphic event to Late Jurassic thrust imbrication of arc units on the Rhodope basement, which is the first time that early Alpine thrust tectonics in the northeastern Rhodope Massif have been recognized.

[4] To whom any correspondence should be addressed.

Acknowledgements
The study was supported by NSFB contract no. VU-NZ 02/06 and SNSF SCOPES grant no. IB7320-111046/1.

Metamorphic style and development of the blueschist- to eclogite-facies rocks, Cyclades, Greece

John C Schumacher[1], **John B Brady**[2], **John T Cheney**[3]
[1]Dept. of Earth Sciences, University of Bristol, Wills Memorial Building, Queen's Road, Bristol BS8 1RJ, UK
[2]Dept. of Geology, Smith College, Northampton MA 01063, USA
[3]Dept. of Geology, Amherst College, Amherst MA 01002, USA

Email: j.c.schumacher@bristol.ac.uk

Abstract. The island of Syros, Greece is part of the Attic-Cycladic blueschist belt, formed during Mesozoic Eurasia-Africa subduction. The rocks of Syros can be broadly divided into three tectono-stratigraphic units: (I) metamorphosed sedimentary and volcanic rocks (marble-schist sequence), (II) remnants of oceanic crust with fault-bounded packages of blueschist/eclogite-facies mafic rocks and serpentinite (mafic-ultramafic rocks) and (III) the Vari gneiss, which is a tectonic klippe. Low-temperature, high-pressure assemblages are found on several islands in the Cyclades. The best preserved of these rocks are on Syros and Sifnos islands. Mineral compositions and peak metamorphic assemblages are similar on both islands. Both islands are considered to share similar P-T histories with highest-pressure mineral assemblages reflecting conditions of at least 15 kbar and about 500°C.

1. Introduction

Constraining the P, T and deformation histories of Syros and Sifnos is a work in progress with many contributors who have applied traditional empirical geothermobarometry, phase equilibria and the pseudosection approach using various mineral assemblages (e.g., recently [1], [2], [3]). Many versions of P-T trajectories for Syros and Sifnos appear in the literature ([2], [4], [5], [6], [7], [8], [3], [9], 10]). Differences among these studies, some major and some minor, can be partly attributed to variation in methods and data that were used, but also to the authors' interpretation of the timing of formation of mineral assemblages, state of compositional preservation of the minerals and coexisting equilibrium assemblage. In spite of the range of published P-T estimates, we believe that a few common and widespread mineral assemblages give good limits on the P-T path that was followed during the subduction and exhumation of the interlayered marbles and schists.

For the marbles and schist on Syros, peak metamorphic conditions are limited to just under 16 kbar at about 480°C and just above 500°C at 15 kbar, which fits well with estimates (15-16 kbar and 500°C) most workers have taken as the peak of metamorphism on both Syros and Sifnos. However, some recent work suggests higher P-T conditions on both Sifnos and Syros. Maximum P-T conditions on Sifnos have been estimated at about 19 kbar and 580 °C ([1]), and [2] got similar results of about 19 kbar and 570 °C for Syros. For Syros, higher peak-metamorphic conditions for the mafic-ultramafic unit proposed by [1] are at about 19 kbar and 525°C, and [3], who applied the geothermobarometer of [11], obtained conditions as high as 19-24 kbar (± 2 kbar) at temperatures of 500-580°C (± 65°). On Syros these higher conditions are problematic, because they exceed P-T limits imposed by assemblages in the marbles. A possible scenario to reconcile these observations is that the mafic and ultramafic rocks, already metamorphosed at higher P-T eclogite-facies conditions, became juxtaposed with slices of the schist and marble units at about 15 kbar and 500°C, which are the peak metamorphic conditions seen in the schist and marble units. The fabrics of the main penetrative deformation would have formed at these condi-

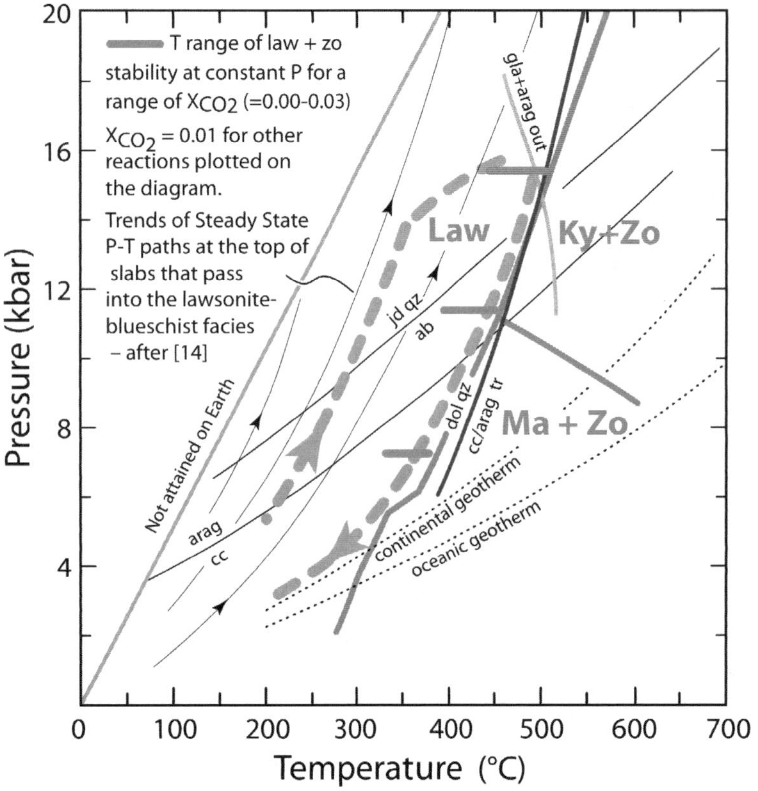

Figure 1. Pressure versus temperature diagram that illustrates the petrologic constraints on the P-T trajectory. Calculations were done with PerPlex [12] using the latest version (HP02) of the thermodynamic data of [13]. Abbreviations: aragonite (arag), albite (ab), calcite (cc), dolomite (dol), glaucophane (gl), jadeite (jd), kyanite (ky), lawsonite (law), margarite (ma), quartz (qz), tremolite (tr) and zoisite (zo).

tions. This explanation would account for the apparently contradictory estimates of peak metamorphic conditions from the marble and schist units and the mafic and ultramafic rocks as well as the observation of many workers (e.g., [8]) that the main deformation occurred at the peak metamorphic conditions.

2. Limits on P, T and grain boundary fluid composition from the schist and marble units

Mineral assemblages that are common across much of Syros provide good constraints on the maximum P-T conditions as well as on the composition of the grain boundary fluid. The assemblage glaucophane + aragonite is stable only at fluid compositions of X_{CO2} = 0.005-0.030. The glaucophane + aragonite/calcite assemblage also constrains the maximum pressure and temperature. The upper pressure boundary is a locus of points that form a curved, concave-downward line with a negative slope. For an X_{CO2} = 0.01 the maximum pressure at which glaucophane + calcite/aragonite could remain stable would be just under 16 kbar at about 480°C (figure 1, orange curve). The maximum temperature would be just above 500 °C at 15 kbar. These estimates bracket the maximum P-T conditions of the schist-marble sequence.

The assemblage dolomite + quartz is also common in many of the low-Na marbles that did not develop sodic amphibole or pyroxene. These marbles did not react to produce either tremolite or diopside, and the implication is that temperatures did not exceed the stability of dolomite + quartz. Breakdown of dolomite + quartz to tremolite (X_{CO2} = 0.01) has a steep P-T slope of about 9-10° per kbar that limits maximum temperature these rocks could have reached to about 475°C at 13 kbar to 560°C at 21 kbar (figure 1).

Within the schists, the preservation of lawsonite also places limits on the P-T conditions, and relict lawsonite + epidote, while difficult to find in hand sample, is relatively common in thin section. Additionally, high-temperature breakdown products of lawsonite, such as kyanite + zoisite (epidote) or margarite + zoisite (epidote), are absent. These two observations suggest that the reactions: lawsonite =

kyanite + zoisite + quartz + H_2O and lawsonite = margarite + zoisite + quartz + H_2O were never crossed and, therefore, limit the maximum T at given P for the peak metamorphism and along the retrograde P-T trajectory (figure 1, gray lines).

Like the glaucophane + aragonite/calcite assemblage in the marbles, lawsonite + epidote in the schists also indicates that water-rich fluids were present for much of the metamorphic history, since calculated T-XCO2 stability of the assemblage lawsonite + zoisite (epidote) for a range of pressures shows that these two phases can only coexist over a narrow range of temperatures and with H_2O-rich fluids (< 3% CO2). Assuming the retrograde P-T trajectory remained largely within the stability field of lawsonite + zoisite, calculations at 15, 11 and 7 kbar place limits on the retrograde P-T trajectory that returned these rocks to the surface (figure 1, green bars).

3. P-T trajectory

Very little evidence to constrain the early prograde metamorphic history of these rocks is preserved. If steady-state P-T paths are a reasonable approximation of the early subduction history of these rocks, then modeling (e.g., [14]) suggests a concave-up prograde paths (figure 1; figure 2, location 1) that pass into the lawsonite-blueschist facies. Pseudomorphs after much of the lawsonite are abundant in the glaucophane schists of Syros show that the subduction P-T trajectory must have entered the lawsonite-blueschist facies, and the rocks must have initially equilibrated there (figure 3a,b). Reaching the conditions of the epidote-blueschist facies, requires a perturbation of the modeled steady-state P-T trajectory, and the essential component is heating at a higher rate (figure 2, location 2). The conditions of the epidote-blueschist facies are marked, partly, by the end of the stability of glaucophane + lawsonite via the reaction glaucophane + lawsonite = clinozoisite + paragonite + chlorite + quartz + water [15]. These conditions are very near the maximum P-T conditions of the marble and schist units, and they mark the beginning lawsonite pseudomorph formation (figure 3a, b). The nature of the lawsonite breakdown reactions is too large a topic to cover here, but other reactions like, lawsonite + celadonite-rich phengite = zoisite + chlorite + muscovite-richer phengite + H_2O are also consistent with these peak P-T conditions and the direction of the P-T trajectory (figure 2, locations 2 & 3) and would also be consistent with some lawsonite pseudomorph formation early in the retrograde history of these rocks [16].

Maximum P-T conditions for the marble and schist units (figure 2, location 3) are based on calculated stability limits of widely-distributed, observed assemblages. The occurrence of glaucophane + $CaCO_3$ + dolomite + quartz suggests that the P-T trajectory that was followed by the rocks crossed a reaction like: albite/Na-pyroxene + dolomite + quartz => glaucophane + $CaCO_3$, but did not exceed the P-T stability of the reaction dolomite + quartz => tremolite + $CaCO_3$. The P-T locations of these reactions are sensitive to fluid composition and indicate that the attending fluid phase was water-rich with X_{CO2} constrained to be < 0.03; a value of $X_{CO2} \cong 0.01$ best fits the observed assemblages. The minerals of these assemblages form the fabric of the major penetrative deformation in the marble-schist sequence. This same fabric is found in the mafic-ultramafic rock unit, so if the mafic-ultramafic rocks follow a different early P-T trajectory, they became juxtaposed with the marble-schist sequence at or slightly before time of the main deformation (partial P-T path with question mark, figure 2). Age determinations from white micas suggest peak metamorphic ages of about 50-55 Ma [10], [17].

Limited retrograde metamorphism began almost immediately after peak temperatures were attained while the rocks were still at blueschist-facies conditions (figure 2, locations 3-4). Where fluids could infiltrate, a common reaction in the omphacite-bearing glaucophane schists was: omphacite + rutile + H_2O = glaucophane + sphene. Assemblages with stable glaucophane + albite + chlorite also form during retrograde metamorphism in the blueschist facies.

The occurrence of relict lawsonite + Al-rich epidote in schists within the marble-schist sequence also places limits on peak and retrograde metamorphic conditions. Calculations show that for pure H_2O fluids, lawsonite + zoisite are stable from about 510 to 450°C, but this range shrinks and terminates at just above 500°C as X_{CO2} approaches 0.03. The preservation of lawsonite and the lack of preservation of

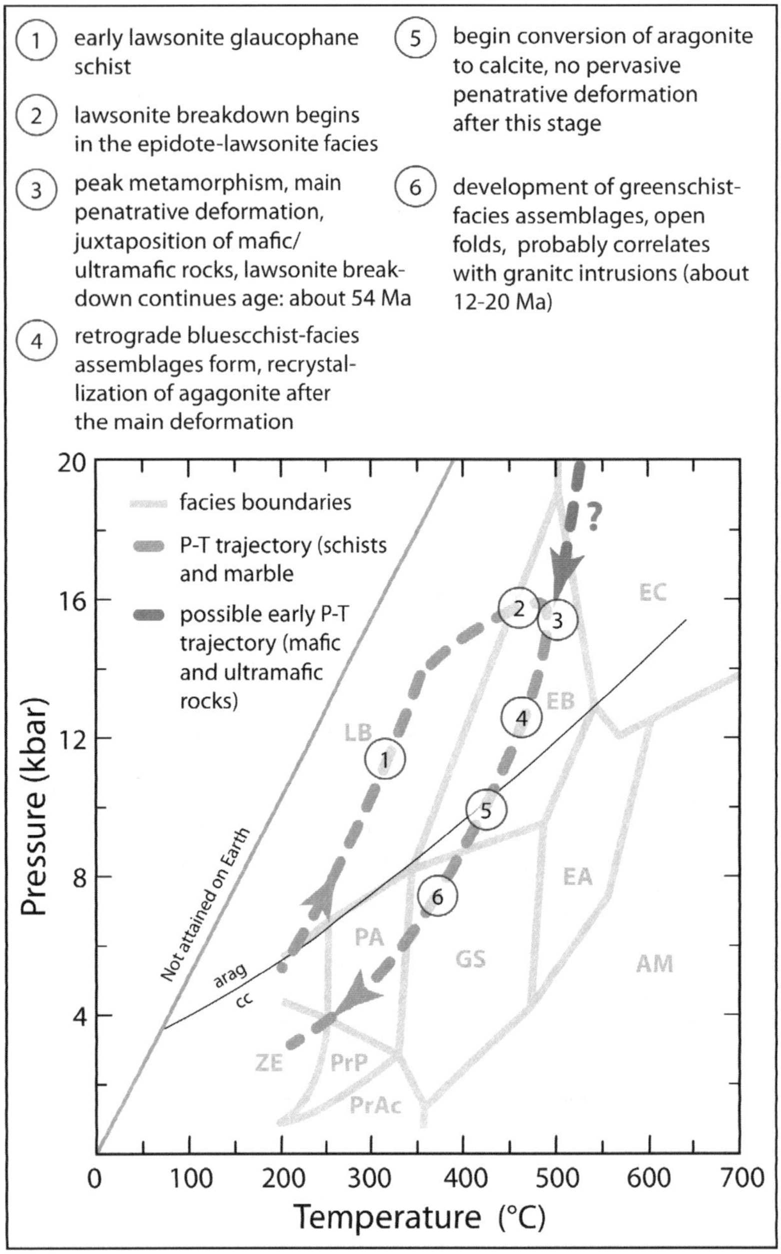

Figure 2. Pressure versus temperature diagram that shows the relative sequence of events along the P-T trajectory. Facies abbreviations: AM=amphibolite; EA= epidote-amphibolite; EB=epidote-blueschist; EC= eclogite; GS=greenschist; LB=lawsonite blueschist; LC=lawsonite-chlorite; PA=pumpellyite-actinolite; PrAc=prehnite-actinolite; PrP=prehnite-pumpellyite; ZE=zeolite [14].

any products of terminal lawsonite breakdown reactions (kyanite or margarite) indicate the P-T trajectory remained within the lawsonite stability field (figure 1, figure 2, locations 4 & 5). Additionally, since lawsonite + Al-rich epidote also have a T-X_{CO2} stability that is limited to fluids with $X_{CO2} < 0.03$, this observation combined with similar fluid compositions indicated by the glaucophane marbles suggests that the whole subduction package of schist, blueschist and marble was pervaded by an H_2O-rich fluid.

The retrograde P-T trajectory (figure 2, location 5) is estimated to have crossed the aragonite to calcite reaction boundary above 400°C, which is consistent with the observed complete conversion of aragonite to calcite, and the formation of the aragonite pseudomorphs. The shape-preferred orientation of the pseudomorphs after aragonite is typically at a high angle to the axial planes of folds (figure 3c), suggesting that aragonite continued to recrystallize after the main penetrative deformation ceased. The

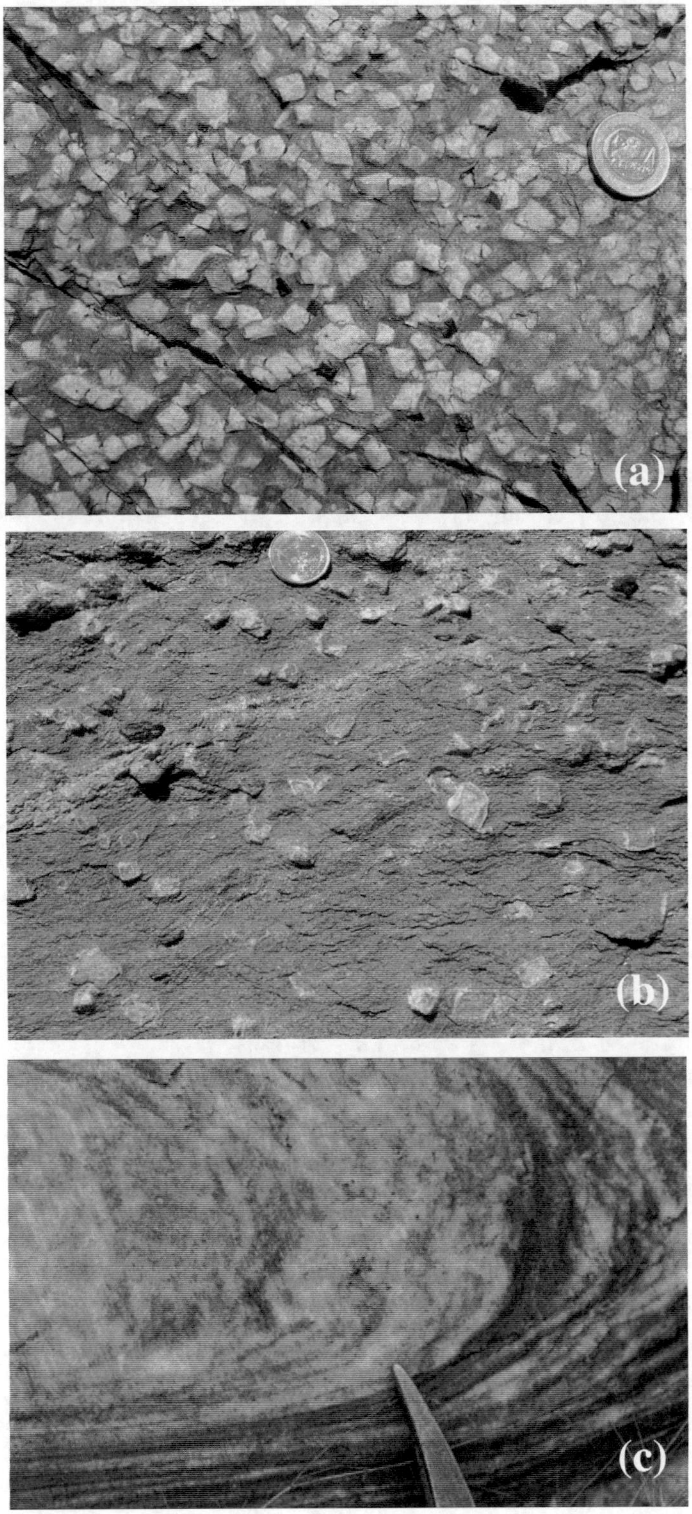

Figure 3. (a) Pseudomorphs (white) after euhedral lawsonite in glaucophane schist from near Grammata in northern Syros. Dark mineral porphyroblasts are coarse-grained rutile. The rock also contains epidote and white mica; (b) Pseudomorphs (white) after euhedral lawsonite in retrograde schist Finikas, southern Syros; (c) Aragonite pseudomorphs (aligned and acicular) showing that aragnite recrystallized after the main fabric forming event. Hammer tip is approximitely oriented parallel to the aragonite pseudomorphs. Sample is from near Vissa, southern Syros.

widespread preservation of the aragonite pseudomorph fabric on Syros indicates that no later penetrative deformation occurred while most of the rocks were within the calcite stability field, since this would have destroyed the observed aragonite pseudomorph fabric[18].

The retrograde P-T trajectory crossed into the greenschist-facies (figure 2, location 6) and much of the observed retrograde mineral growth occurred under these conditions. Elsewhere in the Cyclades greenschist-facies metamorphism is associated with the granitic intrusions (mostly about 20 -12 Ma) [19], and, if this correlation is valid, may roughly constrain the age of this phase of the retrograde metamorphism on Syros.

At low temperatures (<300°C), both the oceanic and continental geotherms converge, and it is reasonable that the P-T trajectory of the marbles and schists would also approach these geotherms near the end of exhumation, since there are no locally observed, late events on Syros that could potentially disturb the geotherm (figure 1; figure 2, location 6).

4. Conclusions

Stability of glaucophane + aragonite/calcite, dolomite + quartz and the presevation of relict lawsonite combined with the absence of minerals formed by terminal lawsonitebreakdown reactions place relatively tight constraints on the P-T conditions for the marble and schist units. Common structural elements indicate that the rocks of the marble and schist unit and the mafic-ultramafic rocks were juxtaposed at least by the time of the main deformation. Higher pressure and temperature estimates may indicate that the mafic-ultramafic rocks followed a different P-T trajectory up to the point they were combined with the marble and schist unit into a single package of rocks. The juxtaposition of this rocks would have occurred at 15-16 kbar at about 500°C, which are the peak metamorphic conditions seen in the marble and schist unit.

5. References

[1] Trotet F, Vidal O, Jolivet L 2001 *European Journal of Mineralogy* **13**, 901-920
[2] Schmädicke E, Will T M 2003 *Journal of Metamorphic Geology* **21**, 799-811
[3] Holly E A , Ross T, Cheney J T 2004 *Geological Society of America Abstr. with Prog*. **36**, A67
[4] Dixon J E, Ridley J 1987 Syros In: H C Helgeson (ed) *Chemical transport in metasomatic processes* p 489-518 D Reidel Publishing Company: Dordrecht
[5] Schliestedt, M (1986) *Journal of Petrology* **27**, 1437-1459
[6] Okrusch M, Bröcker M 1990 *European Journal of Mineralogy* **2**, 451-478
[7] Avigad D, Garfunkel Z 1991 *Tectonophysics* **188**, 357-372
[8] Rosenbaum G, Avigad D, Sanchez-Gomez M 2002 *J. Structural Geology* **24**, 1451-1462
[9] Keiter M, Piepjohn K, Ballhaus C, Bode M, Lagos M 2004 J. *Structural Geology* **26**, 1433-1445
[10] Putlitz B, Cosca M A, Schumacher J C 2005 *Chemical Geology* **214**, 79-98
[11] Ravna, E J K, Terry M P 2004 *J. of Metamorphic Geology* **22**, 579-592
[12] Connolly J A D 2005 *Earth and Planetary Science Letters* **236**, 524-541
[13] Holland, T J B, Powell, R 1998 *Journal of Metamorphic Geology* **16**, 309–343
[14] Peacock, S M 1993 *Geological Society of America Bulletin* **105**, 684-694
[15] Evans, B W 1990 *Lithos* **25**, 3-23
[16] Brady, J B, Able, L M, Cheney, J T, Sperry, A J, and Schumacher, J C 2001 *Geological Society of America Abstr. with Prog*. **33**, 250-251
[17] Bröcker M, Kreuzer H, Matthews A, Okrusch M 1993 *J. Metamorphic Geology* **11**, 223-240
[18] Brady J B, Markley M J, Schumacher J C, Cheney J T, Bianciardi G A 2004 *J. Structural Geology* **26**, 3-9
[19] Andriessen P A M, Boelrijk N A I M, Hebeda E H, Priem H N A, Verdurmen E A T, Vershure R H 1979 *Contributions to Mineralogy and Petrology* **69**, 215-225

Donald D Harrington Symposium on the Geology of the Aegean IOP Publishing
IOP Conf. Series: Earth and Environmental Science **2** (2008) 012018 doi:10.1088/1755-1307/2/1/012018

Diverse P-T paths for the Cycladic ultramafic rock associations: Constraints on their origin, emplacement and exhumation

Yaron Katzir[1,3], **Zvi Garfunkel**[2], **Dov Avigad**[2] **and Alan Matthews**[2]
[1] Dept. of Geological and Environmental Sciences, Ben Gurion University of the Negev, Be'er Sheva 84105, Israel
[2] Institute of Earth Sciences, The Hebrew University of Jerusalem, Jerusalem 91904, Israel

E-mail: ykatzir@bgu.ac.il

Abstract. Within the Alpine orogen of the Cyclades ultramafic rocks occur in diverse tectonic positions, are associated with a variety of rock sequences and are regionally metamorphosed at variable conditions. Here this diversity is interpreted in terms of variable mechanisms for incorporation of mantle rocks into orogenic belts and used to better constrain spatial heterogeneity of P-T burial and exhumation paths in the Cyclades. The Cycladic Blueschist Unit (CBU) was metamorphosed first at high pressure conditions in the Eocene (M1) and variably overprinted during exhumation in the Early Miocene (M2). On the islands of Syros and Evia the CBU includes relatively thin HP-LT ophiolitic mélange intervals hosted by meta-sedimentary sequences. Meta-serpentinites are associated with eclogitized metagabbro and metabasites on Syros, but with epidote blueschists on Evia. Given the wide and comparable geochemical heterogeneity in the Syros and Evian mélanges, the Syros metabasites record higher M1 temperatures (450-500°C) relative to their Evian counterparts (400-430°C). The M2 overprint is manifested in Syros by greenschist facies assemblages (~450°C), whereas pumpellyite-actinolite assemblage (~350°C) overgrew blueschists in Evia. Within the M2 migmatized leucogneiss core of the Naxos dome (~700°C) meta-peridotite lenses are the sole preservers of pre-M2 history. The Naxos meta-peridotites possess relict mantle assemblage, fertile spinel lherzolite chemical composition and mantle-like oxygen isotope ratios. Thus unlike ophiolite-associated ultramafics in the Cyclades that were serpentinized prior to metamorphism, the Naxos metaperidotites were directly emplaced from the subcontinental mantle into an underthrust continent during collision and HP metamorphism. Ultramafic assemblages constrain M1 temperatures on Naxos to 550-650°C. It follows that going southeastwards from Evia through Syros to Naxos progressively deeper levels of the subducted plate are exposed. Correspondingly temperatures of the M2 overprint also increase. The diverse P-T paths of the CBU form an array wherein the deeper a rock sequence is buried, the 'hotter' is its exhumation path. Such a pattern is predicted by thermal modeling of tectonically thickened crust unroofed by either erosion or uniform extension.

1. Back arc extension and exhumation in the Cyclades

The exhumation of the Cycladic blueschists and eclogites from their formation at great depth during

[3] To whom any correspondence should be addressed.

Eocene Alpine compression (M1) to the surface today involved dramatic change in the tectonic setting of the central Aegean region. The thickened orogenic crust collapsed and rocks that were deeply buried during Alpine collision are caught today in an extensional back-arc environment (figure 1). The close coupling between thickening and back arc extension appears to be an efficient mechanism to exhume and preserve high-pressure metamorphic rocks in the Cyclades. [1] questioned this notion and showed that back arc extension in the Aegean lagged behind a significant part of the exhumation of the high-pressure metamorphic rocks. Evidence for initiation of back arc extension in the earliest Miocene is plentiful and includes: (i) beginning of sedimentation in extension-related basins [2]; (ii) Regional extension-controlled granitic plutonism [3]; (iii) activation of normal sense detachments that juxtaposed low pressure rock (Upper tectonic Unit) on top of exhumed high P/T rocks (Lower tectonic Unit, Cycladic Blueschist Unit = CBU) [4]; (IV) Abundance of northwards-dipping ductile extensional fabrics in Early Miocene overprinting assemblages [5].

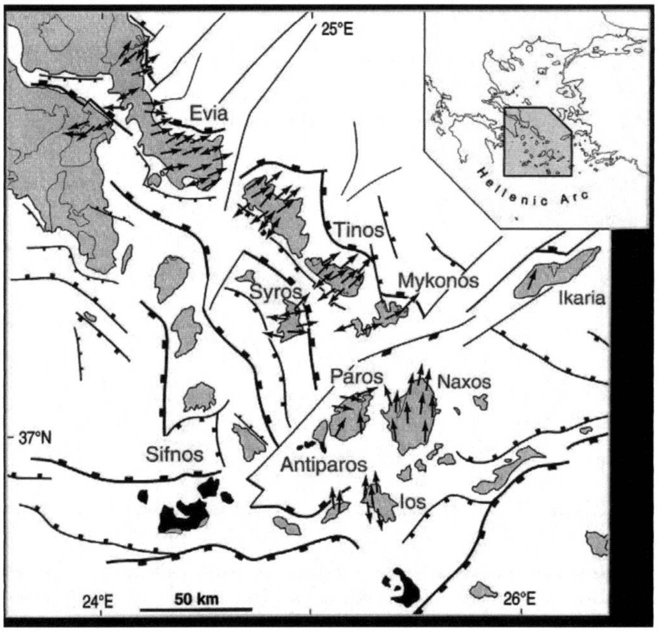

Figure 1. Structural features in the Cyclades including normal (jagged lines) and strike slip (simple lines) faults that have been active since the Pliocene and the sense of ductile shear (arrows) deduced mostly from stretching lineations in greenschist or amphibolite facies minerals within the Lower tectonic Unit (from [5]).

Most of the petrologic studies on the Cycladic high-pressure rocks concentrated on the islands of Sifnos, Syros and Tinos where the best-preserved eclogites occur. The P-T-t paths of rocks from the three islands (the Central Eclogite Axis or CEX) are similar (figure 2): Eocene eclogite facies metamorphism (M1) at 15 ± 3 kbar and 450-500°C was followed by 23-21 Ma greenschist facies overprint at similar temperatures and 5-7 kbar. Hence, the Early Miocene onset of back-arc extension in the Aegean was coeval with greenschist facies metamorphic event (M2) that overprinted high pressure-rocks in the CEX. It is thus evident that at the time of overprinting the Cycladic blueschists and eclogites have already isothermally decompressed from their maximum burial depths to much shallower crustal levels.

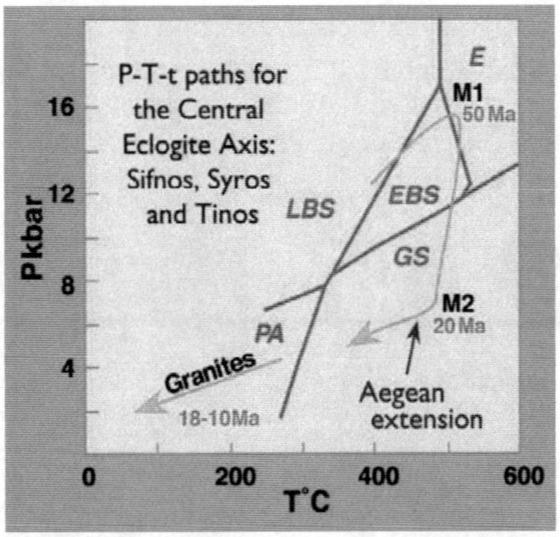

Figure 2. P-T-t path for the Central Eclgoite Axis: Sifnos, Syros and Tinos. Abbreviations: E=eclogite, EBS=epidote blueschist, LBS=lawsonite blueschist, GS=greenschist, PA=pumpellyite actinolite.

2. Naxos - heating during decompression

The M2 metamorphic complex of Naxos forms a N-S elongated, thermal and structural dome (figure 3). The grade of M2 metamorphism increases from low-grade rocks still preserving M1 high-pressure assemblages on SE Naxos to amphibolite-facies rocks in the leucogneiss core of the dome. Peak M2 conditions recorded by migmatites and sillimanite schists in the core are 700°C at 6-8 kbar [6]. Remnants of the older high P/T mineralogy (M1) have not been found in the leucogneiss core of Naxos, thus its exhumation P-T path could not be determined.

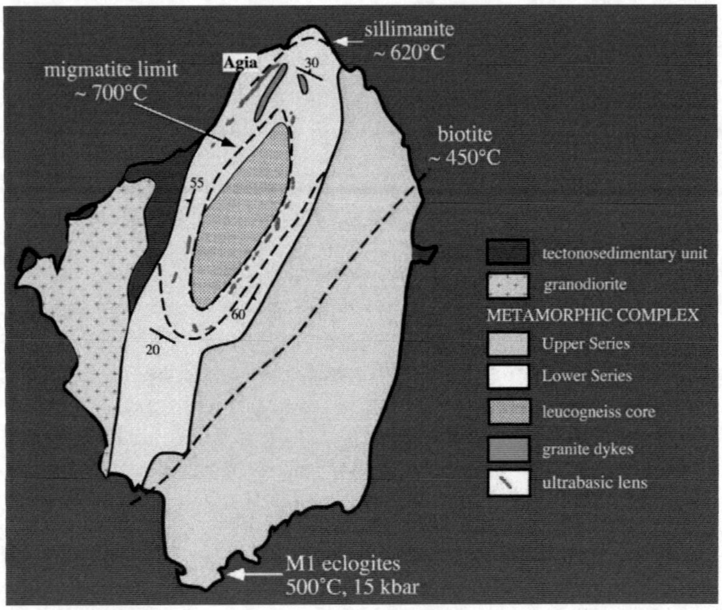

Figure 3. A simplified geological map of Naxos.

First look into the pre-M2 evolution of the Naxos core was made possible by petrologic and oxygen isotope study of peridotite lenses hosted by the silicic gneisses [7, 8]. Structural concordance with the country rocks and metasomatic zonation at the contact with the felsic host rocks indicate that the meta-peridotites have experienced the M2 metamorphism. This conclusion is supported by the similarity between metamorphic temperatures of the ultrabasic rocks and those of the host rocks: maximum temperatures of 730°C were calculated for a syn-kinematic assemblage in the meta-peridotites (figure 4A; Fo-En-Hbl-Chl-Spl).

Figure 4A. M2 high amphibolite facies assemblage in Naxos meta-peridotites: Olivine is the dominant phase, with less abundant OPX and hornblende, all strongly elongated, indicating syn-kinematic recrystallization of the peridotite. Two other components of the same assemblage are chlorite and spinel.

Figure 4B. Orthopyroxene porphyroclast of pre-M2 higher grade peridotite within the M2 metaperidotite groundmass. White scale bar = 0.5mm.

Relics of an older, higher-grade meta-peridotite precursor to the M2 assemblage are well preserved as porphyroclasts of enstatite and forsterite randomly oriented within the M2 recrystallized matrix and showing strong undulose extinction (figure 4B). The pre-kinematic enstatites have high-Al_2O_3 and high-CaO cores (up to 5.5 and 0.9 wt%, respectively) and contain aligned exsolution-lamellae of Cr-spinel. Other relics of the pre-M2 higher-grade peridotite include rare grains of high-Al green spinel (60 wt% Al_2O_3). Aluminium- and Ca-in-Opx thermometry of relict orthopyroxene, olivine and spinel yields temperatures of 1050° ± 20°C. Petrologic analysis thus shows that deep mantle peridotites, formerly equilibrated at the spinel lherzolite field, cooled to at least amphilbolite facies conditions (~600°C) and were then reheated and metamorphosed with the country rocks during M2.

The evolution portrayed by the petrologic data is confirmed by oxygen isotope Opx-Ol thermometry that gives bi-modal distribution of temperatures, grouped at 700° and 1200°C and indicates partial oxygen exchange during M2 superposed over previous mantle fractionation (figure 5).

The occurrence of mantle relicts and the mantle-like chemical and oxygen isotope composition ($\delta^{18}O$(Ol) = 5.2 ‰) indicate that pristine, unaltered peridotites were emplaced into the Naxos section. Since emplacement of peridotites into shallow crustal levels invariably involves serpentinization it is concluded that the peridotites were directly transported from the mantle and tectonically interleaved with the Naxos section at great depth during the Alpine collision and high P/T metamorphism (M1; figure 6). Since Naxos is the only place in the Cyclades where the subcontinental mantle is involved in orogenesis, the Naxos core possibly represents the deepest exposed levels in the Cyclades and thus experienced the highest pressures during collision and M1 metamorphism (600°C, 14 kbar).

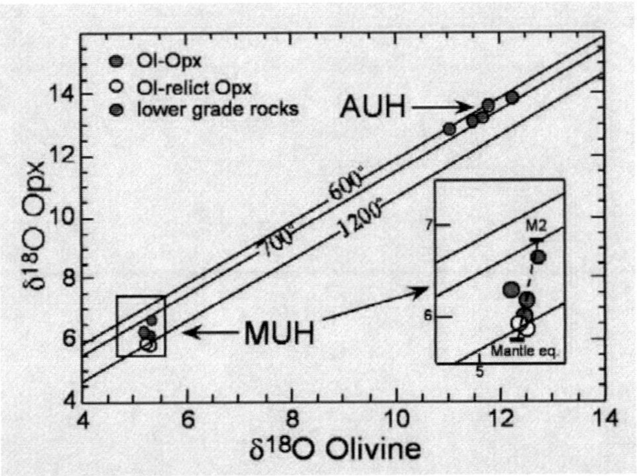

Figure 5. Plot of $\delta^{18}O$ (orthopyroxene) vs. $\delta^{18}O$ (olivine) in Naxos meta-peridotites.

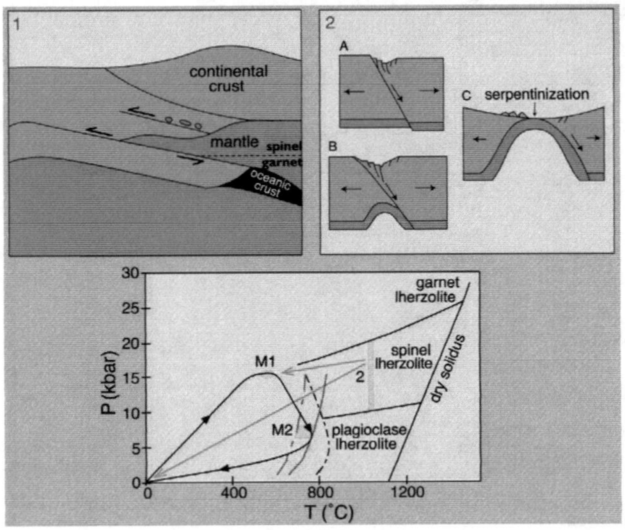

Figure 6. Two possible scenarios for emplacement of meta-peridotites into the Naxos upper continental crust section. (1) Syn collisional emplacement during M1 high pressure metamorphism; (2) Pre-orogenic emplacement of peridotites on the seafloor during rifting. The paths representing the two scenarios (blue arrows) are shown on a P-T diagram also showing estimated conditions for mantle equilibration, M1 and M2 events.

3. Evia - cooling during decompression

The significance of S Evia exposures is that there the allochtonous character of the Lower Unit (Blueschist Unit) is revealed. In the Almyropotamos tectonic window a Basal Unit crops out under the blueschists (figure 7). The Basal Unit consists of platform carbonates containing a Middle Eocene nummulitic hardground at their top and covered by Eocene-Oligocene flysch. The inverted age and metamorphic grade sequence across the Basal-Lower Unit contact indicates that it is a major thrust fault. The thrust operated subsequently to high pressure metamorphism in the Lower Unit and during or later than the accumulation of the flysch in the Late Eocene. Intervals of metamorphosed ophiolitic mélange occur at the upper part of the Blueschist Unit and at its base near the basal thrust.

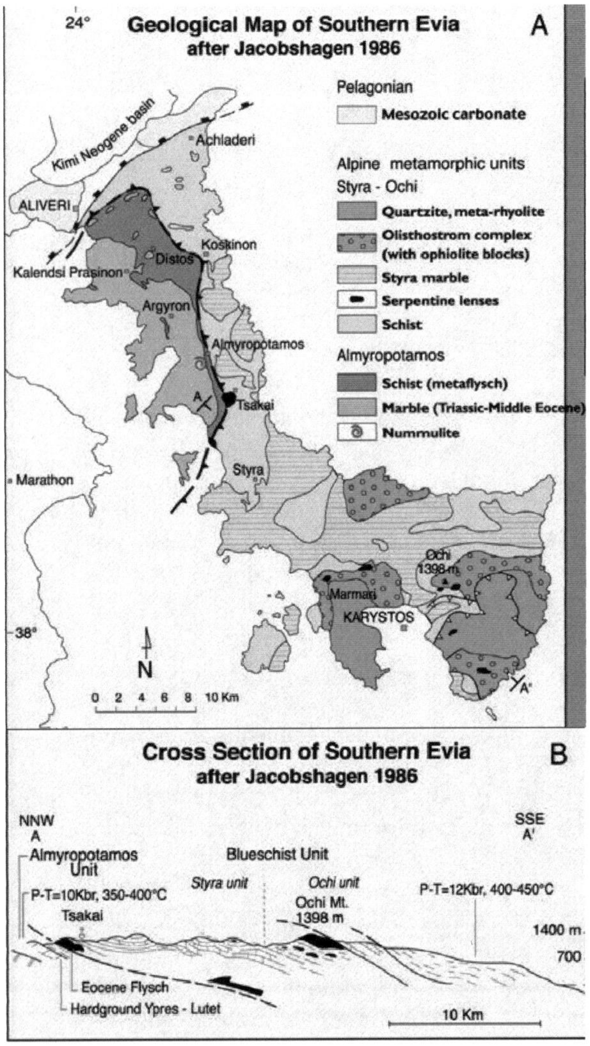

Figure 7A. Geologic map of Southern Evia after [9]. Intervals of metamorphosed ophiolitic mélange occur at the upper part of the Blueschist Unit and at its base near the basal thrust. **B.** Cross section of Southern Evia after [9]. Block of meta-basalt and -gabbro from the Ochi melange preserve glaucophane-epidote-omphacite assemblage that constrains M1 conditions to the epidote blueshist facies field [10].

The metabasites in S Evia were overprinted by pumpellyite-actinolite facies assemblage (figure 8). Pumpellyite crystallized at 300-350°C and 4-6 kbar indicating significant cooling during exhumation.

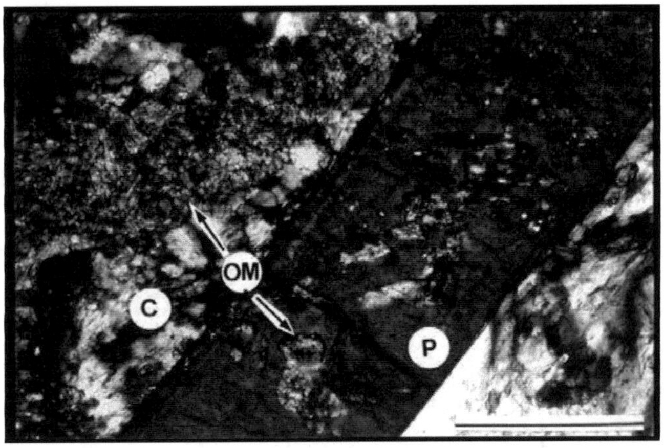

Figure 8. Pumpellyite (P) and chlorite (C) overgow relict omphacite (OM) in a mafic block within the Ochi melange. Scale bar = 0.1 mm

In conclusion, exhumation of the Blueschist Unit in S Evia involved crustal accretion below it, cooling during decompression and an inverted temperature gradient within it. These three features are characteristic of an active accretionary wedge and indicate that substantial part of the exhumation of the Evian blueschists occurred during convergent tectonics.

The high-pressure metamorphic rocks of the Cyclades show variable P-T paths (figure 9). Whereas eclogites on Sifnos, Syros and Tinos decompressed isothermally, eclogites on Naxos were heated and blueschists on Evia cooled during decompression. The pattern formed by the diverse P-T paths is not random: The deeper a rock is buried the 'hotter' is its exhumation path. The reason for this is that the longer a cold rock lingers in deep crustal levels, the more chance it has to approach ambient temperatures. This pattern was long recognized in regions of thickened continental crust unroofed by erosion [11], but also holds if the mechanism responsible for bringing the rocks nearer the surface is uniform extension.

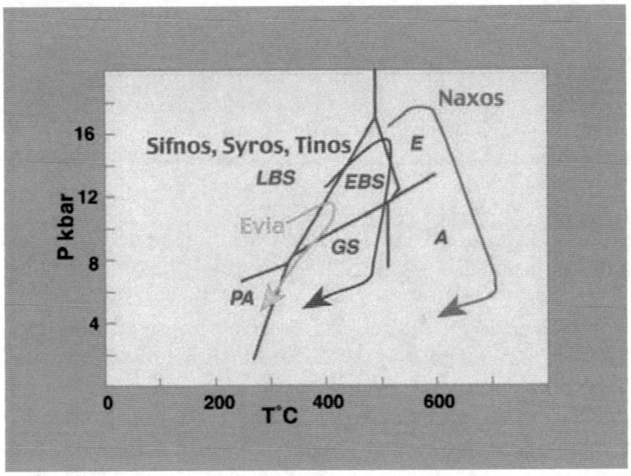

Figure 9. P-T paths for the Cycladic Blueschist Unit in the Central Eclogite axis (blue), on Naxos (orange) and on southern Evia (green) plotted on the metamorphic facies diagram of Evans (1990) [12]. Abbreviations are as in figure 2.

An important phase of unroofing affected the Cycladic eclogites and blueschists between the Eocene and the Early Miocene. Overthrusting of the high-pressure rocks over the Basal Unit took place at the Late Eocene. Either extension or erosion can account for the early stages of exhumation of eclogites both before and after overthrusting. At around 20 Ma whole crust back-arc extension commenced in the Aegean and was accompanied by metamorphic overprint.

4. References

[1] Avigad D, Garfunkel Z, Jolivet L and Azañón J M 1997 *Tectonics* **16** 924–51
[2] Sánchez-Gómez M, Avigad D and Heimann A 2002 *J. Geol. Soc. London* **159** 45–60
[3] Altherr R and Siebel W 2002 *Contrib. Mineral. Petrol.* **143** 397–415
[4] Lister G S, Banga G and Feenstra A 1984 *Geology* **12** 221–25
[5] Gautier P and Brun J P 1994 *Tectonophysics* **238** 399–424
[6] Buick I S and Holland T J B 1989 *Geol. Soc. London Spec. Pub.* **43** 365–69
[7] Katzir Y, Avigad D, Matthews A, Garfunkel Z and Evans B W 1999 *J. Metamorph. Geol.* **17** 301–18
[8] Katzir Y, Valley J W, Matthews A and Spicuzza M J 2002 *Contrib. Mineral. Petrol.* **142** 700–13
[9] Jacobshagen V 1986 *Geologie von Griechenland.* Borntraeger, Berlin.
[10] Katzir Y, Avigad D, Matthews A, Garfunkel Z and Evans B W 2000 *J. Metamorph. Geol.* **18** 699–718
[11] England P C and Richardson S W 1977 *J. Geol. Soc. London* **134** 201–13
[12] Evans B W 1990 *Lithos* **25** 3–23

IOP Conf. Series: Earth and Environmental Science **2** (2008) 012019 doi:10.1088/1755-1307/2/1/012019

The unroofing history of Naxos and Paros: Constraints and questions from thermochronology and thermal modeling

R A Ketcham[1,3] **and S Brichau**[2]

[1] University of Texas at Austin, Jackson School of Geosciences, Geological Science Dept., 1 University Station C1100, Austin, TX 78712-0254, USA

[2] Joint School of Earth Sciences, University and Birkbeck College, Malet Street, London, WC1E 7HX, UK

E-mail: ketcham@geo.utexas.edu

Abstract. The Naxos/Paros extensional fault system (NPEFS) accommodated a significant portion of Miocene extension in the Aegean. The exposure is unusually complete, consisting of a thick lower-crustal ductile shear zone grading upwards into a thin, brittle detachment. Thermochronometric systems with closure temperatures documenting passage of the footwall through the brittle (apatite and zircon fission-track and (U-Th)/He) and ductile (hornblende K-Ar) zones provide important constraints on the unroofing history of this system. Brittle-zone slip rates based on fission-track data are generally in the 6-8 km/Myr range spanning times from 13-8 Ma, whereas inferred ductile-zone slip rates average ~5 km/Myr from 16-12 Ma. We used finite element thermal modeling to test the hypothesis that Naxos and Paros unroofed simultaneously and at similar rates, from ~16 Ma to ~8 Ma, with the only difference being a ~12 Ma granodiorite intrusion on Naxos. We found that a simplified model with uniform detachment fault dip (30°) and slip rate (6.5 km/Myr.) adequately explains all of the fission-track and some zircon (U-Th)/He data, although other (U-Th)/He data are problematic. The data are somewhat more fully reproduced if there was some acceleration during the latter stages of unroofing, but large uncertainties in slip rates preclude a firm conclusion. The thermal model also provides a reasonable mechanism for unifying inferred brittle-zone slip rates with slower ductile-zone ones, as the latter may have been rendered artificially low due to isotherm advance during the early stages of detachment faulting. At the same time, the modeling exercise forces a thorough assessment of the thermal budget of the crust before and during the initiation of detachment faulting. Other data have suggested that unroofing began as early as 20 Ma, and that early decompression was associated with heating. Insofar as unroofing is only in itself a cooling mechanism for the footwall, the source of this additional heat, and what happened to it, is a matter requiring further investigation before a complete picture of extension on the NPEFS can be formulated.

1. Introduction

Physical consideration of large-scale continental extension cannot be divorced from the thermal milieu in which it occurs. Temperature exerts a primary control on both rheology and density, which in turn influence crustal strength, isostasy and uplift. The thermal regime of extension is also the driving force behind hydrothermal processes along detachment faults, influencing the character of deformation and mineralogical transformations observed in its aftermath. Igneous activity is frequently associated with

[3] To whom any correspondence should be addressed.

large-scale extension and detachment faulting, though it remains difficult to determine the extent to which magmatism is a consequence of or contributor to tectonism.

The Aegean region has long been recognized as the locus of a large degree of extension during the Miocene [1], much of which was accommodated by large detachment faults featuring high slip rates. Good exposure and recent activity make this region an ideal natural laboratory for studying the nature of large-scale crustal extension, and the conditions under which it occurs.

This study uses thermochronology combined with thermal modeling to examine the temperature regime of unroofing for the central Aegean islands of Naxos and Paros. A particular point of interest is whether there is evidence of an increase of slip rate along the detachment during transition from ductile to brittle deformation. The exercise of thermal modeling also forces a concurrent examination and evaluation of all of the assumptions that feed into our understanding of this tectonic system, which can then be considered in the broader context of the region.

2. Geological Setting

Naxos and Paros (figure 1) unroofed primarily due to displacement along the Naxos-Paros Extensional Fault System (NPEFS), a top-to-the-north detachment that accommodated roughly 50 km of displacement from ~16 Ma to ~8 Ma. The detachment system grades from a ~1-km-thick ductile zone featuring NNE-SSW stretching lineations in the south to a narrow, brittle contact with the hanging wall in the north. The hanging wall consists of Miocene-Pliocene sedimentary rocks of the Upper unit, whereas the metamorphic cores are dominated by the Cycladic blueschist unit.

Minor S-type granites intruded Naxos and Paros from 15-11 Ma, during motion on the NPEFS. Western Naxos also features a large I-type granodiorite that synkinematically intruded the footwall at ~12 Ma [2,3].

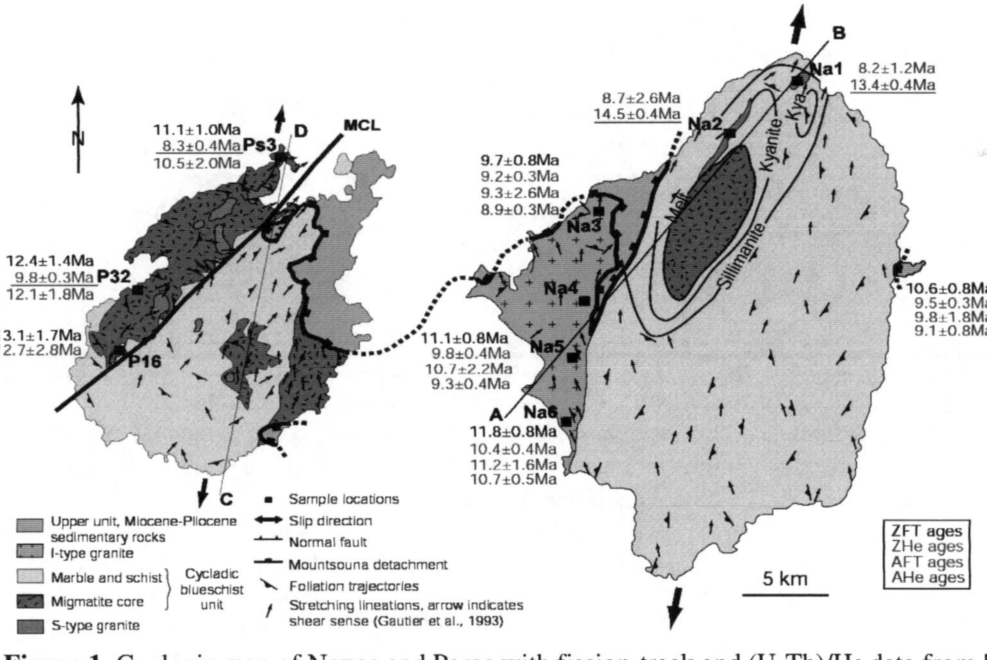

Figure 1. Geologic map of Naxos and Paros with fission-track and (U-Th)/He data from [7].

2.1. Metamorphic conditions

Peak blueschist conditions on Naxos occurred around 45-50 Ma [3]. Well-preserved assemblages in the SE of the island record peak conditions of ~470°C and 12-14 kbar [4], which may be extrapolated to estimate conditions on the structurally lower central island at ~550°C and ~15 kbar, and an overall geothermal gradient of ~13°C/km. The central island experienced a greenschist overprint by ~25 Ma

[2], probably during decompression, which proceeded to a Miocene amphibolite-facies event peaking at 670±50°C, 5-7 kbar at ~16 Ma [5] featuring anatexis leading to a migmatitic core surrounded by onion-pattern Barrovian isograds reflective of a mantled gneiss dome. The geothermal gradient implied by this pattern averages about 30°C/km, 40°C/km near the surface. The SE part of the island decompressed without heating, recording a Miocene overprint at <420°C and 3-5 kbar [4], roughly consistent with being a shallower exposure of extreme conditions in the center of the island.

Central Paros has similar isograd patterns to Naxos, suggesting that exposed rocks reflect only slightly lower Miocene temperatures.

2.2. Prior thermochronologic work

John and Howard [6] compiled K-Ar and Ar-Ar data (hornblende, biotite and white mica) for Naxos from a number of sources to demonstrate a pattern of generally increasing ages from NNE to SSW. Ages fall from ~10 Ma at the detachment contact to ~16 Ma in the SW of the island along a roughly linear trend parallel to the migmatite dome. In the structurally higher SE ages increase to 30-50 Ma. Slip rates for the NPEFS were calculated by plotting distance from the detachment to versus age, resulting in values of 4.4 $^{+3.5}/_{-1.4}$ km/Myr based on hornblende and 5.1-8.6 km/Myr based on biotite.

3. Thermochronology and slip rate determinations

Brichau et al [7] measured apatite and zircon fission-track (AFT, ZFT) and (U-Th)/He (AHe, ZHe) ages on Naxos and Paros to better examine the low-temperature segment of unroofing. Sample localities on Naxos were exclusively syn-extensional intrusives, whereas on Paros the Cycladic basement was sampled. Results are summarized on figure 1.

In a number of localities the ages from 3 or 4 thermochronometers are within their uncertainty bounds of each other, attesting to extremely rapid cooling. Approximate closure temperatures for these systems for fast (100°C/m.y.) cooling are 300°C (ZFT), 220°C (ZHe), 120°C (AFT), and 85°C (AHe), and the partial retention zone for AHe extends down to ~40°C. Thus, the data indicate that the final ~250°C of cooling took place in as little at 1-2 million years.

In general the Brichau data corroborate the spatial pattern of ages observed in earlier work, with ages overall younging from south to north as the detachment zone is approached. There are some inconsistencies in this pattern and among the ages, however. Our expectation was that age determinations from a single locality should decrease with decreasing closure temperature, but there are a number of exceptions, notably the old AHe ages near the detachment contact on Naxos and the young ZHe ages relative to both FT systems on both islands. AHe ages in the S-type granites on Naxos may have been affected by He implantation from the large titanites in those bodies, and thin U-rich rims were observed in zircons during FT analysis, which may have led to young ages due to excessive loss of He to alpha ejection. These are not definitive interpretations, but the data point to the utility of using multiple thermochornometers simultaneously as a means of cross-checking.

Slip rates were calculated from the ages combined with their approximate distances from the detachment contact using χ^2 minimization taking into account both the age uncertainties and 1 km (2σ) uncertainty in distance. Uncertainties in slip rate were estimated using a Monte Carlo procedure, and are reported as asymmetric 95% confidence limits. On Naxos, slip rates calculated for samples Na3-Na6 in km/Myr are: 6.4 $^{+6.8}/_{-2.2}$ (ZFT), 6.5 $^{+15.7}/_{-3.8}$ (AFT), 13.2 $^{+9.4}/_{-4.5}$ (ZHe), and 9.1 $^{+5.3}/_{-2.6}$ (AHe). The AFT rate rises to 8.0 $^{+11.8}/_{-3.0}$ km/Myr if samples Na1 and Na2 are included. Calculated rates for Paros are 6.9 $^{+11.4}/_{-3.1}$ (ZFT), 6.1 $^{+13.7}/_{-3.5}$ (AFT), and 6.1 $^{+2.8}/_{-1.6}$ (ZHe) km/Myr.

The wide confidence intervals stem from the large uncertainties of the individual ages (particularly for FT) and from the limited net displacements (15-25 km) covered by sampling. Uncertainties can be diminished somewhat by combining data for multiple thermochronometers into one slip rate determination. The combined rate for Paros is 6.7 $^{+11.1}/_{-2.8}$ km/Myr. The combined slip rate for Naxos is 10.4 $^{+5.3}/_{-2.7}$ km/Myr, but it is unclear whether combining data with mutual irregularities in this fashion is a good idea. The FT data on Naxos are overall very similar to those on Paros, whereas the (U-Th)/He data are indicative of significantly higher rates.

4. Thermal modeling

To help interpret the thermochronometric data, finite element thermal models of tectonic unroofing were performed using the methods described by [8]. In particular, we tested the hypothesis that Naxos and Paros unroofed simultaneously and at similar rates, with the principal difference being the granodiorite intrusion on Naxos. We also tested whether the slower rates previously determined using higher-temperatures thermochronometers could have been affected by movement of isotherms during rapid cooling.

4.1. Model setup

Aside from the granodiorite intrusion, model parameters were identical for both islands. Unroofing was simulated with a rolling-hinge geometry featuring instantaneous rebound of the footwall during displacement of the hanging wall. The detachment dip was 30°, leveling off at an arbitrary depth of 45 km (i.e. no rocks from this level are exposed today). Due to lack of knowledge about the overall crustal structure in the area, generalized rock properties of granite in the crust and olivine in the mantle were used; heat production was conservatively set at 1 $\mu W/m^3$ for the upper crust and 0.39 $\mu W/m^3$ for the lower crust. Model elements were 2165 m wide and 1250 m deep, and the starting model space was 121 km wide by 120 km deep.

It is usual practice to begin dynamic thermal models from a steady state condition, to isolate the process being studied from transient relaxation effects. In this study, this required setting the mantle heat flow to 50 mW/m^2, an unusually high value, to match the peak metamorphic conditions at 16 Ma of 680°C at a depth of 23 km corresponding approximately to the position of sample Na2 on Naxos. It is almost certain that these conditions reflect a transient event, or a process already underway, but because the natural and geometric expression of this event is unclear modeling any attempt to model it ran the danger of detracting from the less ambitious objectives of this study. Because of the dominant effect of advection and thermal relaxation on cooling it is unlikely that including prior events would have substantively changed modeling results and interpretations, except as noted below.

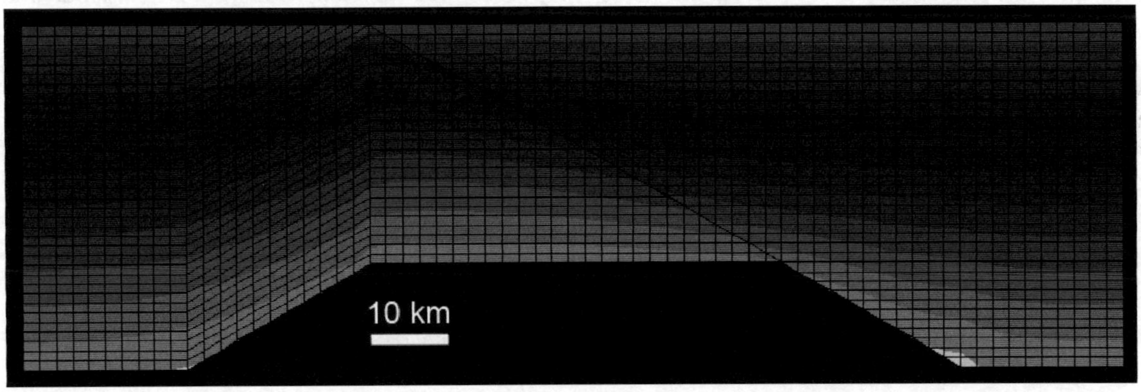

Figure 2. Thermal model of detachment faulting on Naxos during intrusion of granodiorite at 12.3 Ma. Contour interval indicated by colors is 100°C.

The granodiorite on Naxos was simulated by instantaneously resetting all of the nodes it encompassed to 700°C at 12.3 Ma (figure 2). At that stage of the model these nodes lay largely along the detachment surface, extending from 3.8-12.5 km, easily encompassing the conditions implied by thermobarometry. This model was taken as characterizing western Naxos in the area that was intruded, whereas the model without the intrusion was taken to characterize Paros and central Naxos in the area encompassed by previous Ar/Ar work.

4.2. Model results

The modeling was able to simultaneously describe a great deal of data. Using a uniform slip rate of 6.5 km/Myr, all of the FT ages and slip rates for all systems on Paros were replicated to well within estimated uncertainties. Extending these results to central Naxos, it was found that lower slip rates inferred from hornblende Ar/Ar work were replicated as well, as advection of isotherms along the detachment during the early stages of extension served to postpone the closure of early-unroofed specimens relative to late ones. Isotherms achieved a quasi-steady-state by the time the lower-temperature thermochronometers began to close.

Particularly intriguing were the results of the models depicting the intrusion of the granodiorite on western Naxos. Despite the thermal havoc wrought by this event, inferred cooling rates were only slightly affected (0-1.5 km/Myr), and the principal effect on age systematics was to shift them ~2 Ma later. This model was able to replicate all of the FT ages and slip rates using the same 6.5 km/Myr slip rate employed elsewhere, although a slightly better fit to the AFT data for samples Na1 and Na2 was obtained by adding a late acceleration to 8 km/Myr.

5. Discussion and conclusions

The ability of a single pair of models to describe a wide range of data from Naxos and Paros strongly supports the strength of the simple hypothesis of steady, simultaneous unroofing. The model shows that there is no need to call upon an acceleration of slip along the detachment fault from the ductile to brittle zone to explain the observed patterns in the thermochronometric data. However, the modeling does not rule out such an acceleration, either. Furthermore, if detachment faulting began before 16 Ma, then isotherms may already have advected to some extent, possibly rendering this mechanism unable to explain the lower slip rates inferred from the higher-temperature thermochronometers.

It is worth noting that the exercise of modeling also enforced a coherent set of constraints during setup that resulted in a close match to our understanding of the development of these islands: faulting initiating at ~16 Ma and ending at ~8 Ma, intrusion of the granodiorite at ~12 Ma at an appropriate range of depths, and appropriate amounts of heating and displacement. From this standpoint, the modeling served not so much to reveal how unroofing occurred as to encapsulate prior knowledge and assumptions and test whether the new data were consistent with them.

However, important questions on the thermal evolution of these islands remain unanswered, and are in fact highlighted by the modeling. Most significant among these is when extension and unroofing truly began, and what the associated thermal circumstances were. Metamorphic data suggest that central Naxos almost isothermally went from a depth of 45-50 km at 45-50 Ma to ~27 km at 25 Ma, and then heated by ~150°C while rising to a depth of ~23 km by 16 Ma. The source of this heating is of particular interest, as unroofing by any means only serves to cool the footwall, even though the resulting higher transient geothermal gradients give the impression of heating. The observed heating thus requires an additional mechanism beyond those considered in this work.

One possibility suggested by earlier modeling [8] is that the presently exposed footwall actually began in the hanging wall of the detachment system, and was heated by passage of the rising footwall on a more southerly fault. After some amount of unroofing the detachment could have jumped northward to excise a large section of the hanging wall and expose the rocks observed today. This model is exceedingly conjectural, as we know of no evidence of such an earlier detachment, but it may be able to explain the observations without resorting to a mechanism fundamentally different from those already known to be in operation.

A second possibility is that a massive, unseen intrusion, perhaps in the form of a sill, underlies the islands and is responsible for the metamorphic patterns. This may be even more conjectural, as the scale of intrusion necessary to produce these effects on such a large spatial scale is quite massive. For example, even the granodiorite that dominates western Naxos has a contact aureole that extends only ~1-2 km from its margin, whereas the gneiss dome and associated Barrovian isograds associated with the heating in question are 10's of km in some directions and cover an area of 100's of square kilometers. Although this may be possible in the context of a thermal aureole augmented by fluid convection, the symmetry and extent of alteration on Naxos and its replication on Paros appear to require very fortuitous circumstances. In general, a crust-scale effect requires a crust-scale process.

A third and perhaps related possibility lies in the "space problem" associated with large-scale crustal extension. The means by which the lower crust and upper mantle accommodate extension in these environments remains unclear. It is possible that mantle upwelling, lower crustal flow, and rising of anatectic melts, individually or in any combination, could add a great deal of thermal energy to extensional system. However, it should be noted that such a thermal effect appears to be absent from the core complexes in the southern Basin and Range of the United States, perhaps signifying a fundamental difference in the evolution of these otherwise similar structural and tectonic domains.

References

[1] McKenzie D 1978 Active tectonics of the Alpine-Himalayan belt: the Aegean Sea and surrounding regions *Geophys. J. Roy. Astr. S.* **55** 217–54

[2] Andriessen P A M Boelrijk N A I M Hebeda E H Priem E H Verdurmen T Verschure R H 1979 Dating the events of metamorphism and granitic magmatism in the Alpine Orogen of Naxos (Cyclades, Greece) *Contrib. Mineral. Petrol.* **69** 215–25

[3] Wijbrans J R McDougall I 1988 Metamorphic evolution of the Attic Cycladic metamorphic belt on Naxos (Cyclades, Greece) utilizing $^{40}Ar/^{39}Ar$ age spectrum measurements *J. Metamorph. Geol.* **6** 571–94

[4] Avigad D 1998 High-pressure metamorphism and cooling on SE Naxos (Cyclades, Greece) *Eur. J. Mineral.* **10** 1309–19

[5] Buick I S Holland T J B 1989 The P-T-t path associated with crustal extension, Naxos, Cyclades, Greece *Evolution of Metamorphic Belts (Geological Society Special Publication* No 43*)* ed J S Daly R A Cliff B W D Yardley (London: Geological Society) pp 365–9

[6] John B E Howard K A 1995 Rapid extension recorded by cooling-age patterns and brittle deformation, Naxos, Greece *J. Geophys. Res.* **100** 9969–79

[7] Brichau S Ring U Ketcham R A Carter A Stockli D Brunel M 2006 Constraining the long-term evolution of the slip rate for a major extensional fault system in the central Aegean, Greece, usng thermochronology *Earth. Planet. Sci. Lett.* **241** 293–306

[8] Ketcham R A 1996 Thermal models of core complex evolution in Arizona and New Guinea: implications for ancient cooling paths and present-day heat flow *Tectonics* **15** 933–51

IOP Conf. Series: Earth and Environmental Science **2** (2008) 012020 doi:10.1088/1755-1307/2/1/012020

Evidence for polymetamorphic garnet growth in the Çine (southern Menderes) Massif, Western Turkey

Courteney B Baker[1], **Elizabeth J Catlos**[1,2,5], **Sorena S Sorensen**[3], **Ibrahim Çemen**[1] **and Mete Hancer**[4]

[1] Oklahoma State University, School of Geology, 105 Noble Research Center, Stillwater, OK, 74078, USA.

[2] University of Texas at Austin, Jackson School of Geosciences, Geological Science Dept., 1 University Station C1100, Austin, TX 78712-0254, USA

[3] Smithsonian Institution Museum of Natural History, PO Box 37012, MRC 119, Washington, DC 20013-7012, USA

[4] Pamukkale Universitesi, Muhendislik Fakultesi Jeoloji Muh. Bolumu, Denizli, 20070, Turkey

E-mail: b.baker@okstate.edu, ejcatlos@gmail.com

Abstract. Garnet-based thermobarometry is often used to develop models for the evolution of the Menderes Massif, a key Aegean metamorphic core complex. Here we present X-ray element maps and high-contrast backscattered electron (BSE) and cathodoluminescence (CL) images from a garnet-bearing rock from the Çine (southern Menderes) Massif. The images document a polymetamorphic history as plagioclase and garnet grains show distinct cores and rims. The sample contains matrix monazite in reaction with allanite. The garnet in the sample is likely not in equilibrium with its matrix minerals. This is evidenced by BSE images that document compositional variability in both core and rim zoning and tracks of bright streaks extending from rim to core. We propose that some garnet that is now present in the Menderes Massif formed due to collision during Cambro-Ordovician and may have recrystallized during subsequent collisional and extensional events. These processes led to non-equilibrium compositions and can result in spurious pressure-temperature (*P-T*) calculations. To establish the feasibility of the *P-T* estimates of rocks from the Çine Massif for input into tectonic models for the region, more than one sample from single outcrops should be analyzed. Rocks within the Çine Massif have been suggested to display inverted metamorphism, an increase in *T* towards structurally higher levels. Based on the garnet documented here, we propose that the inverted metamorphism may be a consequence of apparent *P-T* rather than a real phenomenon.

1. Introduction

At some point in the Cenozoic, the Aegean region experienced a switch from compressional to extensional tectonics [see review by 1]. Due to a large exposure of high-grade metamorphic and igneous assemblages, the Menderes Massif in western Turkey (figure 1) is an ideal region to study this transition. The massif's complex tectonic history has been of interest to many geologists since the 1960s [e.g., 2]. The Menderes Massif is a large topographic feature (40,000 km^2) and is considered a metamorphic core complex [e.g., 3, 4].

[5] To whom any correspondence should be addressed.

The Menderes Massif is only one of several domains in the Aegean region that experienced the transition from compressional to extensional tectonics, notably the Rhodope, Kazdag, Cycladic, and Crete Massifs (figure 1). All Aegean massifs share a similar lithology, being composed of high-grade metamorphic and igneous rocks, typically bounded by low-angle extensional structures that were active in the late Cenozoic [5, 6, 7, 8, 9, 10]. Numerous researchers have explored the nature, style, and formation of the Menderes Massif [e.g., 4, 8, 11, 12, 13, 14, 15].

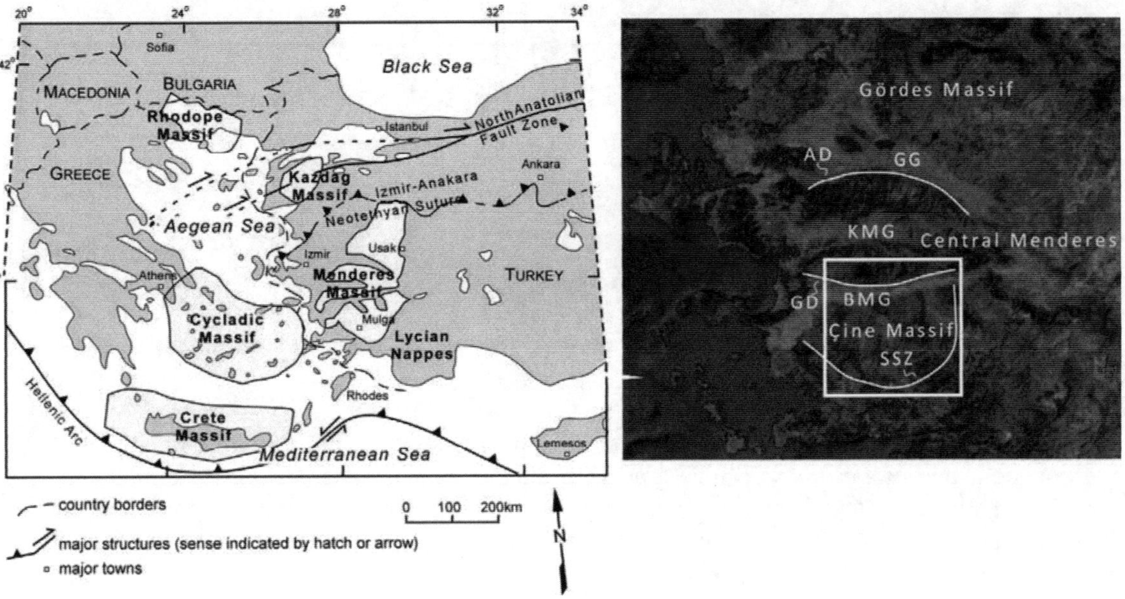

Figure 1. (Left) Generalized map of the Aegean region [after 16]. (Right) NASA WorldWind image of the Menderes Massif with major structural features indicated. Abbreviations: AD= Alasehir Detachment; GG= Gediz Graben, KD= Kuzey Detachment.

Garnet-based thermobarometry have been used to develop models for the evolution of the Menderes Massif [e.g., 12, 17]. Pressure-temperature (*P-T*) conditions are key for ascertaining the mechanisms by which extension created the massif [e.g., 18]. However, geochronologic data from the region show the evidence of multiple tectonic events, from the Cambro-Ordovician to the early/late Cenozoic [Cambro-Ordovician: 19-24; Eocene-Oligocene: 19, 20, 25, 26; Pliocene: 16]. The timing of extensional exhumation is controversial, but is largely thought to have initiated sometime in the mid- to late Cenozoic (Oligocene-Miocene) [4, 13, 26, 27, 28].

In this paper, we present backscattered electron images and X-ray element maps from a garnet-bearing rock (sample EB23) collected from the southern portion of the Menderes Massif (Çine Massif) (figure 2). The possibility of multiple stages of garnet growth in the region has implications on previous thermobarometric methods, and impacts our understanding of models for the development of the Menderes Massif.

2. Geological Background

The Menderes Massif is divided into northern (Gördes), central, and southern (Çine) submassifs based on the presence of E-W trending grabens: the northern Alaşehir (Gediz) and southern Büyük Menderes (figures 1 and 2) [see 29]. These grabens separate the central Menderes metamorphic core complex from the adjacent Gördes Massif to the north and the Çine Massif to the south.

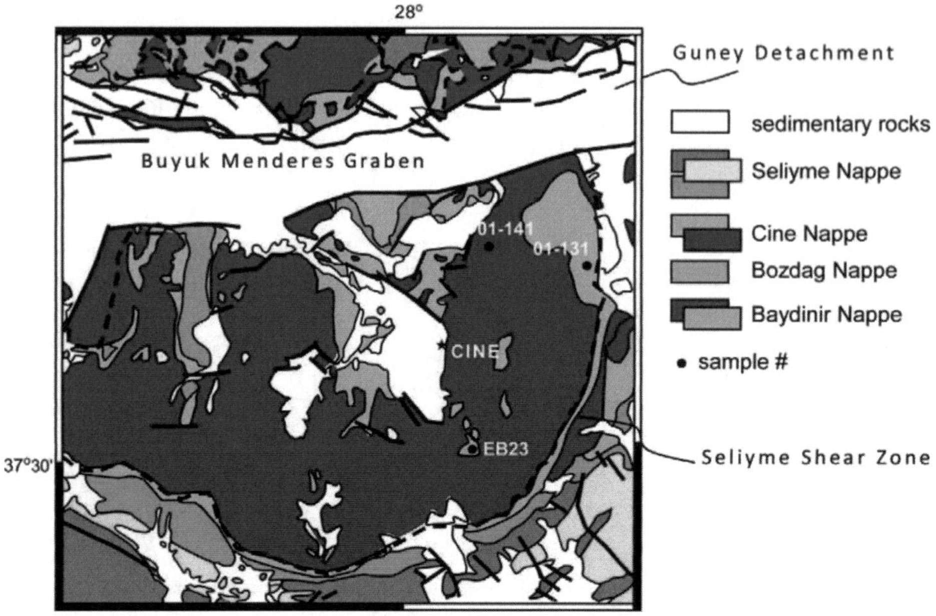

Figure 2. Geologic map of the Çine Massif after [30] and our observations. Samples 01-141 and 01-131 after [16].

Rocks of the Menderes Massif are classified by lithology and metamorphic history [e.g., 4]. In one interpretation, the lowest unit is a high-grade Cambro-Ordovician gneissic series termed the "Core Series" and the overlying Paleozoic schist and Mesozoic-Cenozoic marble are considered "Cover Series" [e.g., 11, 19, 22, 31, 32]. The Core Series consists of granitic augen gneisses, paragneisses, schists, and minor migmatites. The oldest clearly documented geological event in the Core Series is the intrusion of large volumes of granites during the Cambrian in the southern and central Menderes massif [20, 22]. These intrusions are presumably related to a Pan-African orogenic event [31, 32]. The boundary between the core and cover has been debated, and has been thought to be an unconformity [2, 22, 31], extensional shear zone [e.g., 33], incipient detachment along which granite intruded [e.g., 34], a southward facing thrust [e.g., 4, 8, 15], or a thrust overprinted by extensional movement [e.g., 26, 35].

The traditional core-cover interpretation has been further interpreted, and the Menderes Massif has been divided into a series of nappes. From the base to the top, these nappes are termed the: Bayindir, Bozdag, Çine, and Seliyme nappes (figure 2) [see 4, 8, 16, 36]. The Bayindir nappe is structurally lowest, and consists of shelf sediments metamorphosed to lower greenschist faces [4] during the Cenozoic. This metamorphism, often referred to as the "Main Menderes Metamorphism" [e.g., 14, 37] is commonly assumed to be Paleocene-Eocene and to be pervasive throughout the massif. The precise age of this Barrovian metamorphism is not directly constrained, but Rb-Sr whole rock and $^{40}Ar/^{39}Ar$ ages imply Eocene or older [19, 20, 26].

In the overlying Çine and Bozdag nappes (figures 2 and 3), Precambrian deformation and metamorphism are evident from eclogites (600-650 Ma), migmatites (540-550 Ma), and post-deformational plutons (540-560 Ma) [e.g., 19, 22, 24]. The Çine nappe is largely comprised of amphibolite to granulite facies ortho- and paragneisses intercalated with metabasite [4, 38], whereas the Bozdag nappe is characterized by amphibolite-facies garnet-mica schists. These units are interpreted as Pan-African in origin, but with a Cenozoic greenschist [4] to amphibolite-grade metamorphic overprint [e.g., 16].

The Seliyme nappe, Cycladic blueschists, and Lycian nappes overlie the Çine and Bozdag nappes, and contain high-P, low-T assemblages (e.g., Mg-carpholite) [14, 39]. The Seliyme nappe is a metasedimentary sequence with a Precambrian basal section overprinted by movement during the Alpine [4].

The Çine and Bozdag nappes show an inverted metamorphic gradient [4, 40]. Garnet-bearing assemblages in the Çine nappe record peak P-T conditions of 670°C-730°C and 6.2-6.3 kbar, whereas Bozdag rocks located beneath the Çine nappe reached 480°C-540°C and 6.1-7.6 kbar [40]. To ascertain if garnet-bearing assemblages of the Çine nappe record a polymetamorphic history, we obtained X-ray element maps and high contrast backscattered electron (BSE) and CL images from a Çine nappe sample (EB23) (figure 2).

3. Results

Sample EB23 was collected from the Çine nappe in the Çine Massif (figure 2) and contains Grt + Bt + Ms + Chl + Pl + Ilm + Zrn + Qtz + Mnz + Aln + Ap; abbreviations after [41]. Sample 01-141 (figure 2), located nearby has a similar assemblage, and contains garnets with inclusions of monazite that yield ages of 501±18 Ma, whereas matrix grains average 42±5 Ma [16]. Garnets in sample 01-131 (figure 2) shows evidence of a sharp decrease in Ca and increase in Mn at their outer rims. These ages and zoning patters raise the possibility that the garnets Çine nappe may have grown at different stages and at different times [16].

High contrast BSE images of sample EB23 show that garnets in this rock have distinct cores and rims (figure 3). The core of the garnet in EB23 is idioblastic and is partly surrounded by a darker mid rim and a brighter outer rim. The BSE image shows tracks of lighter material that extend from the outer rim to the mid-rim, which may be remnants of pathways for regressive fluids. The core of the garnet is distinct in that it contains higher amounts of Y than the mid-rim area (figure 3).

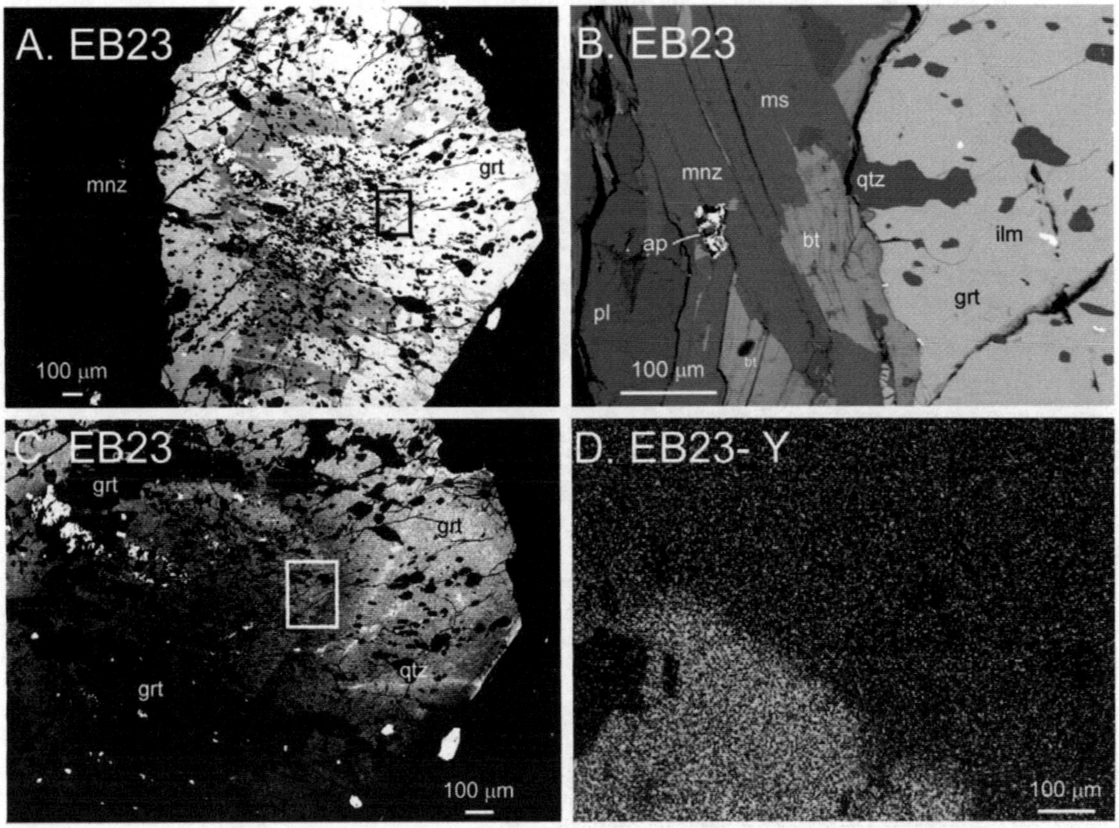

Figure 3. (A-C) BSE images of garnet in Cine Massif sample EB23 (see figure 1 for sample location). The same monazite is indicated in both images (A) and (B). (D) Y-element map of the area in the garnet outlined in (A) and (C). Abbreviations after [41]. Streaks in (C) are not the outcome of sample charging.

Monazite in this rock is present in reaction with apatite, and appears poikiloblastic with radiating inclusions (figures 3 and 4). The textures are consistent hydrothermal alteration [42]. Attempts were made to date these grains using the Th-Pb ion microprobe method [16], but high amounts of common Pb precluded meaningful results. Spots on the grains only yielded only 38% to 3% radiogenic $^{208}Pb^*$.

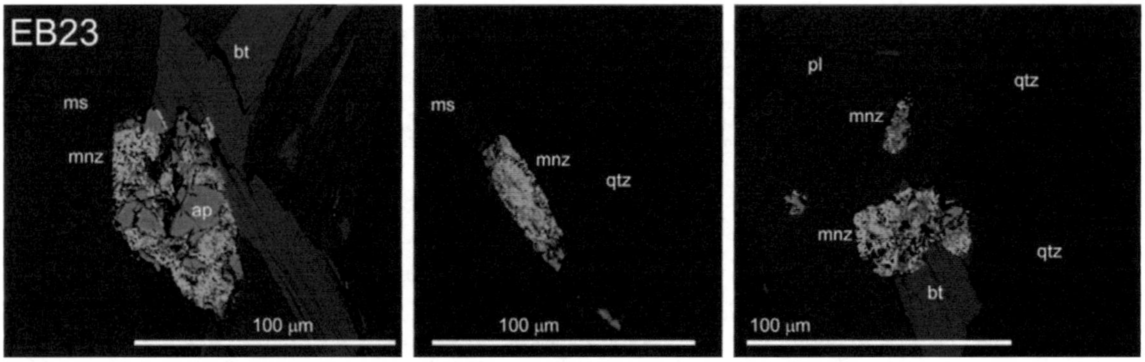

Figure 4. BSE images of monazite in sample EB23 (see figure 1 for sample location).

The rock was imaged using CL techniques using a Premier American Technologies luminoscope model ELM-3R located in the Smithsonian Institution's Department of Mineral Sciences at the National Museum of Natural History [for methods, see 43]. Although the garnet does not luminesce, CL images of the garnet are as useful as the BSE images in identifying distinct cores and rims (figure 5) as they can be used to document regions of high inclusion density and highlight inclusions that vary in luminescence.

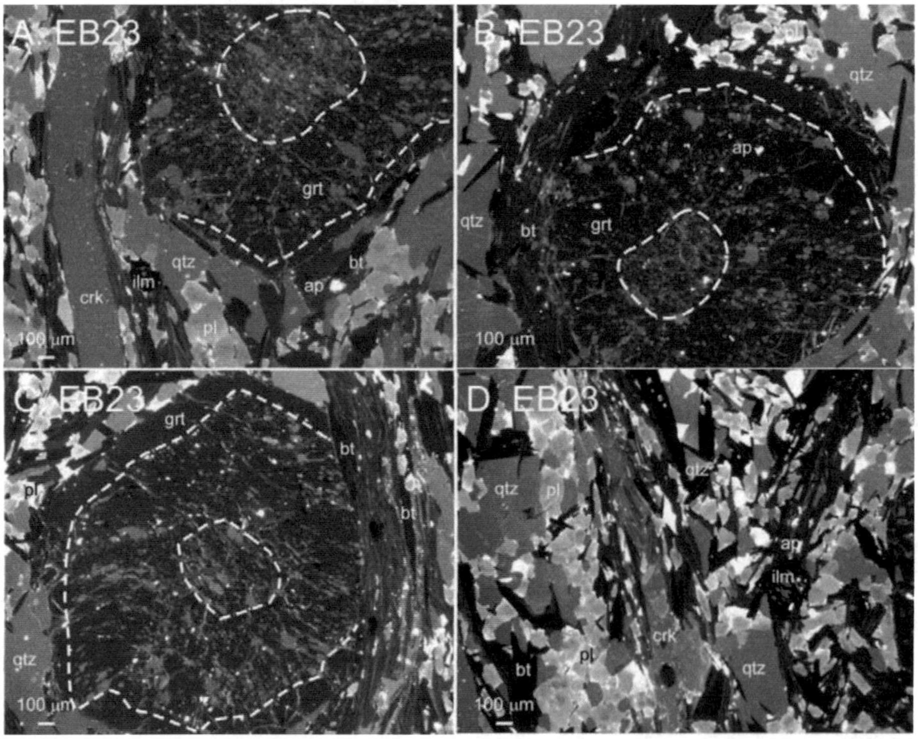

Figure 5. (A-C) CL images of garnets and (D) rock matrix in sample EB23. The garnet in image (A) is the same as in figure 2.

We identify 3 distinct regions in the garnet: (1) an inclusion filled core area that may be high in Y (figure 3), (2) A mid-rim area that contains larger (100mm-sized) xenoblastic quartz inclusions and cracks that appear to radiate from the core, and (3) an inclusion free outer rim. Although cracks are present in the outer rim area (see also figure 2), some terminate at the mid-rim area.

Quartz in the rock luminesces red-brown (figure 5) and we did not see evidence for deformation lamellae or changes in color from core to rim. Some quartz grains appear brighter than others (figure 5D), but this may be the result of their proximity to cracks in the thin section. Plagioclase in the rock has red brown cores, rims of purple and light purple, and white outer rims. Grain boundaries between the plagioclase grains are distinct, as many are outlined in white.

4. Conclusions

High contrast BSE and CL images of sample EB23 from the Çine nappe of the Çine Massif document evidence of multiple stages of garnet growth. The images challenge the view that the nappe experienced "a single Barrovian-type metamorphism" [17] and that garnets in the Menderes Massif formed during a single time [35, 44]. Inverted metamorphism between the Çine and overlying Bozdag nappes may be an outcome of the stacking of nappes that experienced multiple stages of metamorphism. Menderes Massif garnets may have grown during the Cambro-Ordovician, Eocene-Oligocene, or Miocene; note that monazite inclusions in Menderes Massif garnets are Cambrian to Eocene [16].

Acknowledgments

This work was supported by an award to E.J Catlos and I. Çemen from the National Science Foundation (NSF- 0440169). We thank Kitty Milliken at the UT Austin Electron Microprobe Laboratory. The sample was collected with assistance from M. J. Kohn, M. Cemal Göncüoglu, and E. Diniz. We appreciate the comments by M. J. Kohn that improved the original version.

5. References

[1] Çemen I, Catlos E J, Gogus O and Ozerdem C 2006 *GSA Special Paper* **409** 353–79
[2] Schuiling R D 1962 *Bull. Res. Expl. Inst. Turkey* **58** 71–84
[3] Dewey J F 1988 *Tectonics* **7** 1123–39
[4] Gessner K, Ring U, Passchier C W, Johnson C, Hetzel R and Güngör T 2001 *Geology* **29** 611–14
[5] Buick I S 1991 *J. Geol. Soc. London* **148** 93–103
[6] Katzir Y, Avigad A, Matthew Z, Garfunkel Z and Evans B W 1999 *J. Metamorph. Geol.* **17** 301–18
[7] Finger F, Krenn E, Riegler G, Romano S and Zulauf G 2002 *Terra Nova* **14** 233–40
[8] Ring U, Gessner K, Güngör T and Passchier, C W 1999 *J. Geol. Soc. London* **156** 3–6
[9] Bonev N 2006 *GSA Special Paper* **409** 211–35
[10] Fassoulas C, Kilias A and Mountrakis D 1994 *Tectonics* **13** 125–38
[11] Şengör A M C and Yilmaz Y 1981 *Tectonophysics* **75** 181–241
[12] Ashworth J R and Evirgen M M 1985 *Min. Mag.* **121** 323–37
[13] Seyitoğlu G and Scott B C 1996 *J. of Geodyn.* **22** 145–53
[14] Rimmele G, Oberhaensli R, Goffe B, Jolivet L, Candan O and Çetinkaplan M 2003 *Lithos* **71** 19–46
[15] Regnier J L, Ring U, Passchier C W, Gessner K and Güngör T 2003 *J. Metamorph. Geol.* **21** 699–721
[16] Catlos E J and Çemen I 2005 *Int. J. Earth Sci.* **94** 204–17
[17] Regnier J, Mezger J E and Passchier C W 2007 *Geol. Mag.* **144** 67–104
[18] Bozkurt E and Oberhaensli R 2001 *Int. J. Earth Sci.* **89** 679–708
[19] Satır M, and Friedrichsen H 1986 *Int. J. Earth Sci.* **75** 703–14
[20] Hetzel R, and Reischmann T 1996 *Geol. Mag.* **133** 505–72

[21] Hetzel R, Romer R L, Candan O and Passchier C W 1998 *Geologische Rundschau* **87** 394–406
[22] Loos S and Reischmann T 1999 *J. Geol. Soc. London* **156** 1021–30
[23] Candan O, Dora O O, Oberhansli R, Centinkaplan M, Partzach J H, Warkus F C, and Durr S 2001 *Int. J. Earth Sci.* **89** 793–811
[24] Gessner K, Collins A S, Ring U and Güngör T 2004 *J. Geol. Soc.* **161** 93–101
[25] Bozkurt E and Satir M 2000 *Geol. J.* **35** 285–96
[26] Lips A L W, Cassard D, Sözbilir H and Yılmaz H 2001 *Int. J. Earth Sci.* **89** 781–92
[27] Jolivet L, Daniel J M, Truffert C and Goffe B 1994 *Lithos* **33** 3–30
[28] Hetzel R, Passchier C W, Ring U and Dora O O 1995 *Geology* **23** 455–58
[29] Bozkurt E, Park R G, Loos S and Reischmann, T 2001 *J. Geol. Soc. London* **158** 393–95
[30] Konak N 2002 *General Directorate of Mineral Research and Exploration. Izmir Area Map*
[31] Şengör A M C, Satir M, and Akkok R 1984 *Tectonics* **3** 693–707
[32] Dora O O, Candan O, Kaya O, Durr S and Oberhaensli R 1995 *Proceedings of International Earth Sciences Colloquium on the Aegean Region*, 4–14 October 1995, Izmir/Güllük-Turkey, 53–72
[33] Bozkurt E and Park R G 1994 *J. Geol. Soc. London* **151** 213–16
[34] Bozkurt E, Winchester J A and Park R G 1995 *Geol. Mag.* **132** 287–301
[35] Whitney D L and Bozkurt E 2002 *GSA Bulletin* **114** 829–38
[36] Isik V and Tekeli O 2001 *Int. J. Earth Sci.* **89** 757–65
[37] Erdoğan B and Güngör T 2004 *Turkish J. Earth Sci.* **13** 15–36
[38] Sengun F, Candan O, Dora O O and Koralay O E 2006 *Turkish J. Earth Sci.* **15** 321–42
[39] Oberhaensli R, Partzsch J, Candan O and Cetinkaplan M 2001 *Int. J. Earth Sci.* **89** 867–73
[40] Ring U, Willner A P and Lackmann W 2001 *Am. J. Sci.* **301** 912–44
[41] Kretz R 1983 *Am. Min.* **68** 277–79
[42] Poitrasson F, Chenery S R and Bland D J 1996 *Earth Planet. Sci. Lett.* **145** 79–96
[43] Sorensen S, Harlow G and Rumble D 2006 *Am. Min.* **91** 979–96
[44] Bozkurt E 1996 *Turkish J. Earth Sci.* **5** 105–21

Geoarchaeological Research of the Mid-Age Ilyas Bey Complex Buildings with Ground Penetrating Radar in Miletus, Aydin, Western Anatolia, Turkey

Selma Kadıoğlu[1,4], Yusuf Kagan Kadıoğlu[2] and Ali Akin Akyol[3]
[1] Ankara University, Eng. Fac. Geophysical Engineering Dept., 06100 Ankara, Turkey
[2]Ankara University, Eng. Fac. Geological Engineering Dept., 06100 Ankara, Turkey
[3]Ankara University, Baskent Vac. High Sch. Prog. of Restoration and Cons., Ankara, Turkey

E-mail: kadioglu@eng.ankara.edu.tr, kadi@eng.ankara.edu.tr, akyol@ankara.edu.tr

Abstract. The ancient Miletus which were one of the most important city of ancient Iona, are today of great value from cultural standpoint of Turkey. Miletus, situated near the village of Balat in the present district of Soke was founded on a peninsula, approximately 2.5 km long. In the Byzantine period, the city boundaries were quite reduced. In 1424 Miletus was taken inside of the Ottoman Empire and was completely abandoned in the 17th century. Ancient Miletus excavation studies were first begun in 1899 by in Berlin Museum and interrupted during the World War I. At present, the extensive restoration works in Ilyas Bey Complex has applied as a project since 2006. Ilyas Bey Complex that includes Mosque, Medresah and baths situated on the archaeological area in ancient Miletus. Impressive Mosque built in 1404 by Ilyas Bey, Emir of Menteseogullari founded in 1279 and the complex was named after him, is one of the most remarkable buildings of mid-age Miletus. There are two main purposes of the study are (1) to determine archaeological remains of the study area underneath Ilyas Bey Complex and (2) to define the nature of main rock unit and their sources in the vicinity or Aegean region. After preliminary archaeometrical studies, acquired GPR profile data paralleled each other in Ilyas Bey Mosque and its around, Medresah Courtyard and inner Courtyard of the Mosque. After processing 2D parallel GPR profiles, we constructed 3D data volume by lining processed 2D profiles up to correlate remain signatures from each profile for each studied area. It was obtained transparent 3D visualisation of GPR data by assigning a new colour scale for the amplitude range and by constructing a new opacity function instead of the linear opacity function. Therefore we could successfully image the archaeological remains in an interactive transparent 3D volume and its sub-volumes, starting at different depth levels or limited profiles. The archaeometrical (geological and mineralogical, petrographical) studies reveal that the main bulding rock units of Ilyas Bey Mosque are mainly composed of 4 main units these are marble, metalimestone, mica gneiss and granodiorite. The marble, metalimestone with the mica gneiss were obtained from Menderes Massif in Western Anatolia. The granodiorite columns might be obtained from Kozak Pluton in the north of Bergama-Izmir line of western Anatolia.

1. Introduction

[4] To whom any correspondence should be addressed.

Miletus, modern Balat (Soke, Aydın, TURKEY), is about eighty km away from Izmir (figure 1). It is on a hillside surrounded by large ports at the mouth of the river Maeandros at the head of the gulf of Latmus. Now the sea is far off, silted up by the river. Not only ancient Miletus had trade links with countries on the Mediterranean, the Black Sea and the Sea of Marmara but also mid-age Miletus is at the one of the end point of Silk Road route from China on this coast. Miletus was of the most powerful and important Greek metropolises in Asia Minor and was part of the influential Ionian Confederation. The city encouraged erudition and founded several schools attracting scientists (Thales), architects (Hippodamus), philosophers and geographers. Western philosophy originated in ancient Miletus and it is the native city of famous men such as Anaximenes, Hippodamus, Cadmus and Isidorus. The first coins were minted in this city, exact weight measures were established and the Phoenician alphabet perfected. From the results of excavations, it appears that Miletus was actually the Hittite settlement Millawanda and it was certainly a Mycenaean colony by the fourteenth century B.C. Miletus then fell into Persian hands and razed to the ground. It was rebuilt in 479 B.C. by Hippodamus, the famous architect and due course passed to the Romans. It began to prosper once more, new streets and monuments were constructed and it was renamed Castrum Palatia. The Apostle Paul preached Christianity here and the Emperor Justinianus built city walls. The Venetians established commercial trading posts. In 1424, under Sultan Murat II, Miletus became a city of the Ottoman Empire but as the port silted up caused by detritus brought down by the river Maeandros, the city's decline began.

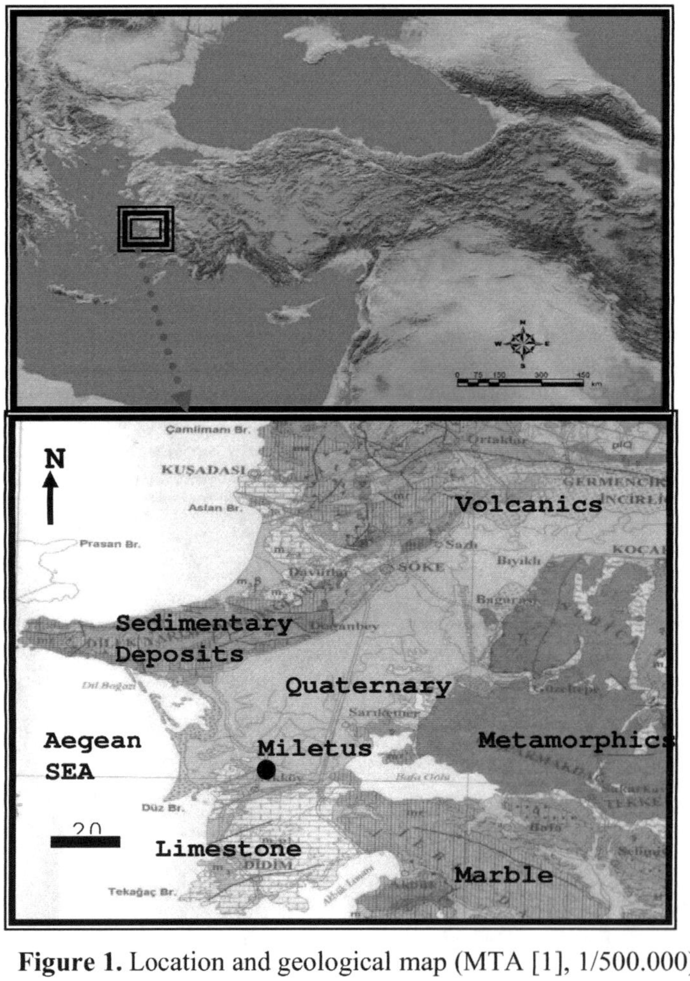

Figure 1. Location and geological map (MTA [1], 1/500.000) of the Ilyas Bey Mosque.

The mid-age Ilyas Bey Mosque near ancient south agora is an exquisite mosque, the Ilyas Bey Mosque, built in 1404 A.D. in the reign of Ilyas Bey. The mosque has a square floor plan surmounted by a central dome but the minaret has collapsed. The decorations above the entrance door and the niche are highly elaborate.

The main purposes of this study are to determine archaeological remains of the study area and to define nature of the main rock unit and their sources in the vicinity of Aegean region.

2. Petrography

The samples were collected for the petrographical determinations and to define their physical properties. The thin sections of all the collected samples of the complex building stone (rock) and mortar with the plaster are investigated under the polarizing microscope. The rocks of the building stones are mainly represented by marble, schist, limestone, travertine and granodiorite. The main compositions and physical properties of the collected samples from Ilyas Bey Complex are given in Tables 1 and 2. The microphotographs of the collected rock samples in the vicinity of Ilyas Bey Mosque are presented in figure 2.

Table 1. Petrographical determinations and physical properties of Ilyas Bey Complex rocks.

Sample Code	Hardness (Mohs)	Density (g/cm^3)	Rock Name	Nitrite (NO_2^-)*	Nitrate (NO_3^-)	Phospahte (PO_4^{3-})	pH	WSS** (%)
MKM-T1	3	2.65	Micritic Limestone	0.150	25	0.200	7.98	1.92
MKM-T2	3	2.66	Marble	0.075	25	0.200	8.25	1.78
MKM-T3	7	2.67	Ms Schist	0.075	25	0	7.78	0.54
MKM-T4	7	2.67	Granodiorite	0.100	25	0.200	7.74	1.70
MKM-T5	3	2.66	Marble	0.100	10	0.200	7.34	0.65
MKM-T6	5	2.67	Chl + Ms schist	0.050	25	0.200	7.17	3.48
MKM-T7	3	2.63	Travertine	0.025	25	0	7.12	3.67
MKM-T8	7	2.67	Qtz mica schist	0.025	25	0.200	7.31	2.50
MKM-T9	7	2.68	Qtz monzonite	0.100	25	0.400	7.22	1.76
MKM-T10	6	2.66	Bt calcschist	0.050	25	0.200	7.36	1.12
MKM-T11	3	2.65	Marble	0.100	10	0	7.68	2.58
MKM-T12	7	2.67	Chl + Ms schist	0.100	25	0.200	7.14	2.25
MKM-T13	3	2.66	Marble	0.100	25	0	7.25	0.99
MKM-T14	3	2.66	Fine crystalline marble	0.200	25	0	6.99	1.72
MKM-T15	3	2.68	Meta granite	0.100	25	0	6.87	2.68
MKM-T16	3	2.67	Fine crystalline marble					2.70

(*) Spot Test of Soluble Salt Determination (mg/L); (**) Total Water Soluble Salt Determination (%)
Mineral abbreviations: Ms, Muscovite; Chl, chlorite; Qtz, quartz; Bt, biotite.

Table 2. Acidic aggregate and binder analysis and aggregate granulometry of plaster and mortar.

Sample Code	Total Carbonate (%)	Total Aggregate (%)	Percent Aggregate Particle Size Distribution by Systematically Sieving					
			<63 µm	63-125 µm	125-250 µm	250-500 µm	500-1000 µm	>1000 µm
MKM-M1	61.02	38.98	0.81	1.71	6.34	21.92	25.11	44.11
MKM-M2	19.07	80.93	1.08	4.61	12.32	19.02	13.57	49.39
MKM-M3	41.40	58.60	1.66	5.57	7.75	14.65	19.04	51.33
MKM-M4	34.19	65.81	1.44	4.64	12.24	19.77	15.29	46.62
MKM-M5	79.06	20.94	2.49	3.68	7.65	15.83	19.89	50.45
MKM-M7	55.82	44.18	1.22	2.39	6.60	20.75	23.84	45.20
MKM-M8	60.32	39.68	0.86	1.52	5.77	20.08	24.20	47.57
MKM-P1	96.73	3.27	10.39	18.33	19.37	21.84	11.58	18.48
MKM-P2	95.65	4.35	6.55	9.35	12.99	16.66	17.99	36.46
M Ave.	**50.13**	**49.87**	**1.37**	**3.45**	**8.38**	**18.86**	**20.13**	**47.81**
Pl Ave.	**96.19**	**3.81**	**8.47**	**13.84**	**16.18**	**19.25**	**14.79**	**27.47**

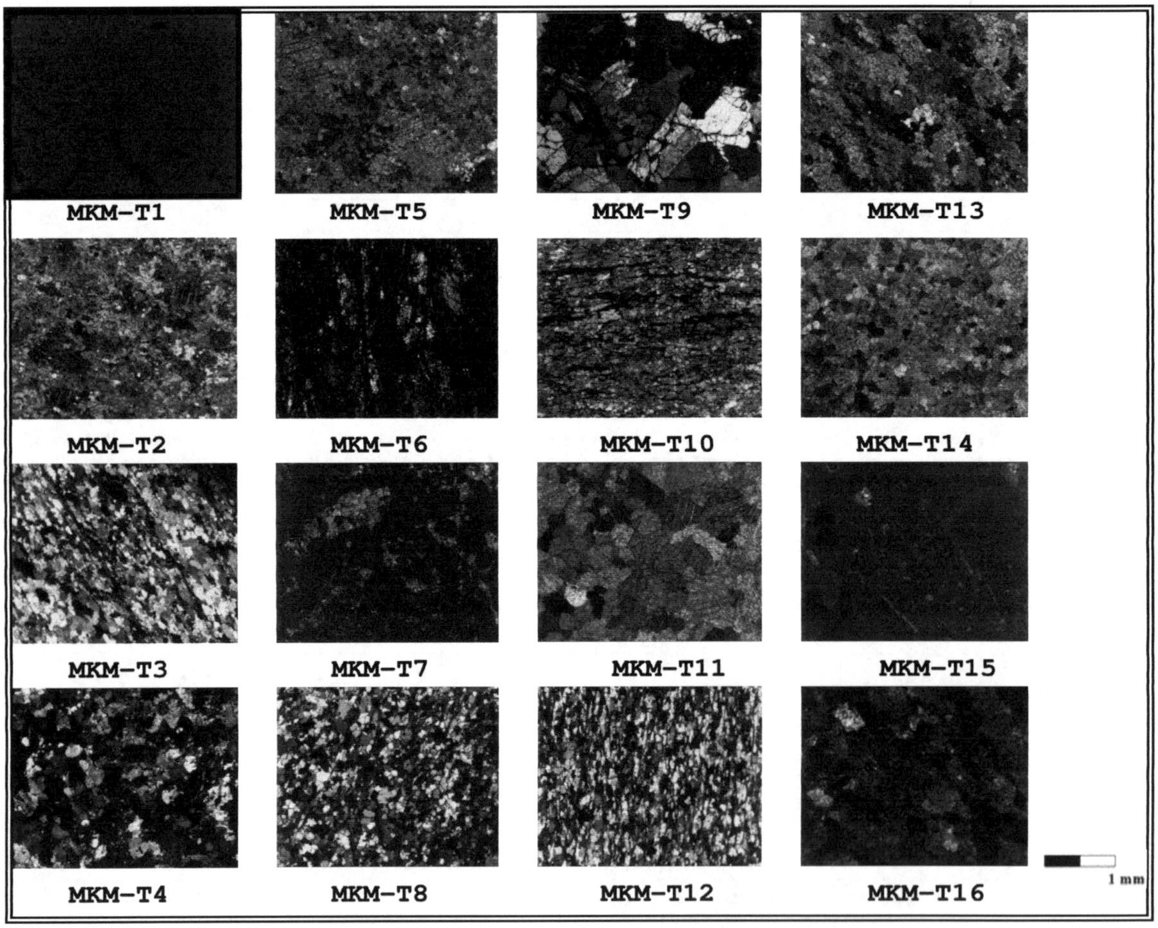

Figure 2. Microphotographs of the rocks of the Ilyas Bey Mosque

IOP Conf. Series: Earth and Environmental Science **2** (2008) 012021 doi:10.1088/1755-1307/2/1/012021

3. Ground Penetrating Radar

Ground penetrating radar (GPR) method is a high frequency electromagnetic geophysical technique commonly used for environmental, engineering, archaeological and other shallow investigations. GPR method uses the scattered wave field of the high frequency electromagnetic (EM) waves to determine the buried archaeological remains. The remains can be identified by the two dimensional (2D) and three dimensional (3D) GPR data visualisation with approximate shape and depth [2, 3, 4].

The acquired GPR profile was established as paralleled each other with 1 m apart in Ilyas Bey Mosque and its behind, left and right side, Courtyard and inner Courtyard of the Mosque (figure 3) by using 500 MHz shielded antenna with RAMAC CU II (Mala Geoscience) system. Figure 4 indicates the plan Ilyas Bey Mosque and paralleled GPR profiles.

WinEKKO-Pro software (Sensor and Software, Canada) package was used for pre-processing of the profiles and GPHYZGPR [5] program in IDL 5.5 was used for data processing as well as obtaining transparent GPR data blocks. The pre-processing was included equalization of the trace number of the profiles and a removing the zero delay time. We constructed 3D volume of the parallel combined 2D GPR data. Low frequency effects and high frequency noises were removed from signals with a second order band-pass Butterworth frequency filter. Amplitude decay compensation, with the same linear function, was applied to all traces of the profiles. Velocity analysis was realized on different diffraction hyperbolas on the different profiles. This analysis yielded mean velocity of 0.1 m/ns for the study area. The time scale transformed to the depth scale by using this mean velocity. Average trace subtraction which was equivalent to F-K filtering for zero degrees of dip [5] was applied to all profiles to diminish horizontal or small degrees of dip noises such as effect of slope which located on the south side of the study area. Finally migration processing was applied to 2D GPR data using mean velocity.

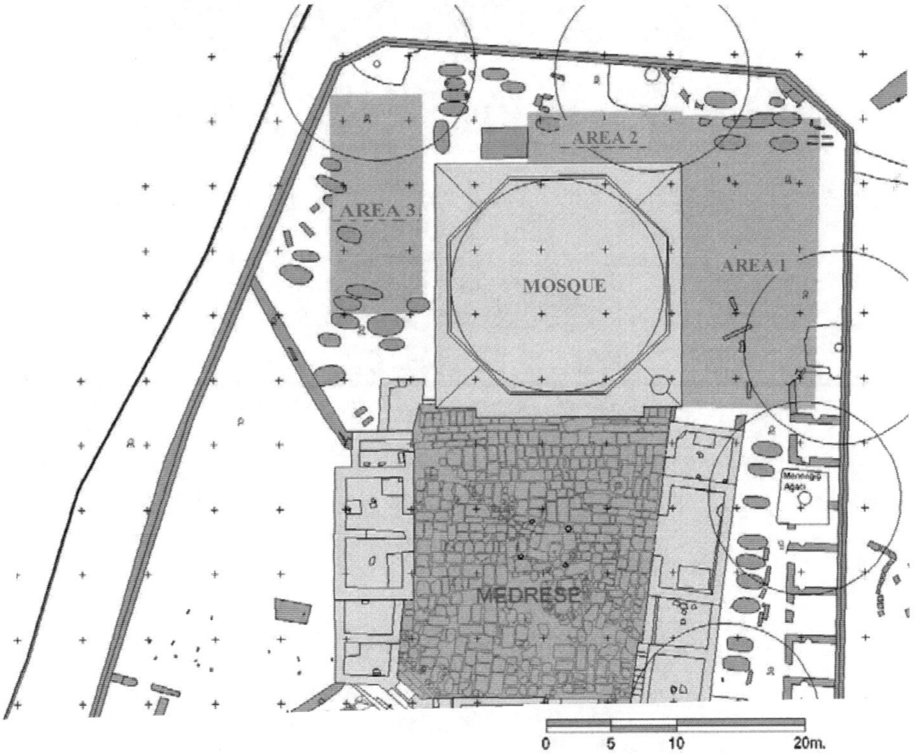

Figure 3. GPR studies in Ilyas Bey Mosque and its behind, left and right side, inner Courtyard of the Mosque (Medrese).

The transparent 3D visualisation of parallel constructed GPR profile data was obtained by assigning a new colour scale for the amplitude range and by constructing a new opacity function instead of the linear opacity function [2, 3]. Therefore we could successfully image the archaeological remains in an interactive transparent 3D volume and its sub-volumes (pillows), starting at different depth levels or limited profiles (figures 5, 6 and 7). We showed some results in Ilyas Bey Mosque to indicate fractures with two processed GPR profile (figure 5) and wall structures determined under the Mosque with two depth slices (figure 6A) and two depth pillows (figure 6B). Figure 7A indicates Medresah plan and GPR profiles with 1 m apart. We determined a buried cistern and related wall structures in the Medresah and photographed with two profile pillows (figure 7B). The horizontal x-axis indicates the distance along the measuring profile. The horizontal y-axis represents the profile sequence. The vertical axis indicates depth from the surface into the underground in figures 6 and 7B.

A

B

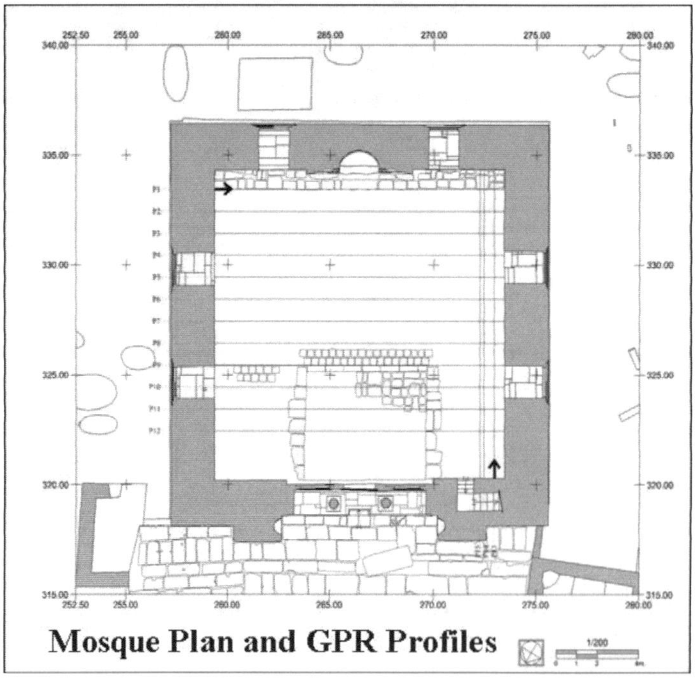

Figure 4. (A) Ilyas Bey Mosque and its inner Courtyard (Medresah), **(B)** Mosque plan and measured GPR profiles in the Mosque (P means profile; P1 indicates profile 1).

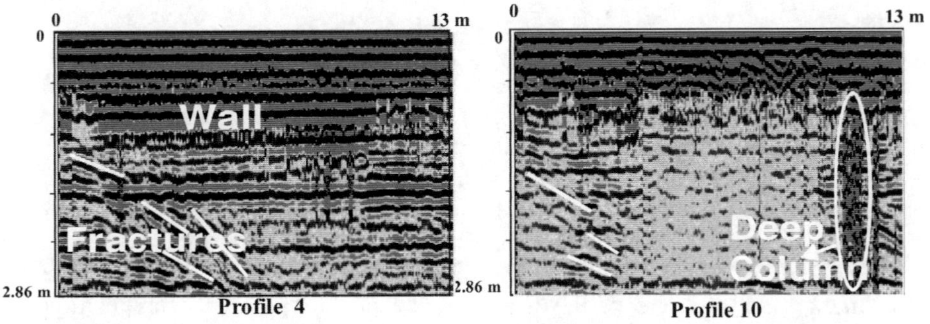

Figure 5. Processed and interpreted GPR profile section 4 and Profile section 10.

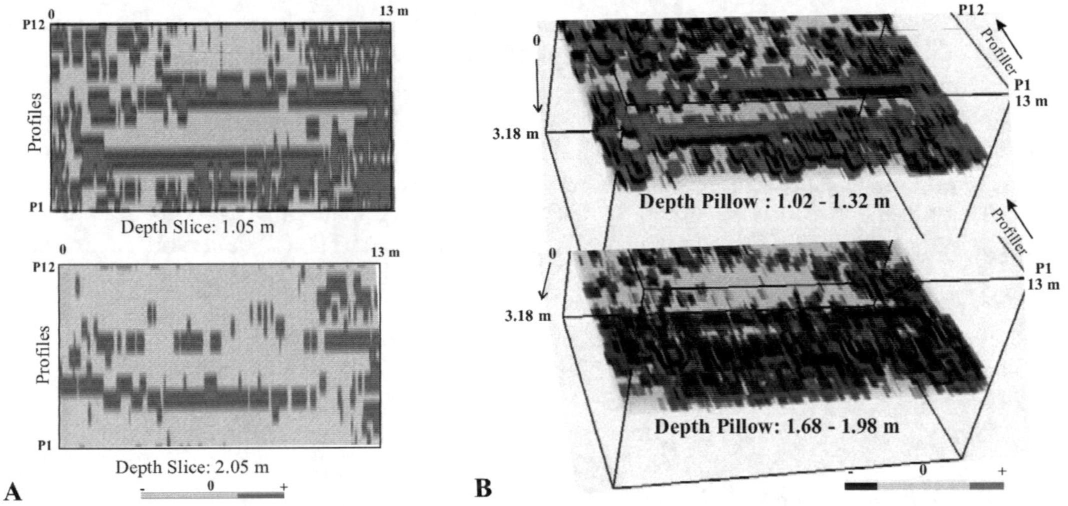

Figure 6. (A) GPR depth slices at 1.05m and 2.05 m **(B)** Depth pillows of the 3D GPR visualization between 1.02-1.32; and 1.68-1.98. Red and black colours indicate wall structures under the Mosque.

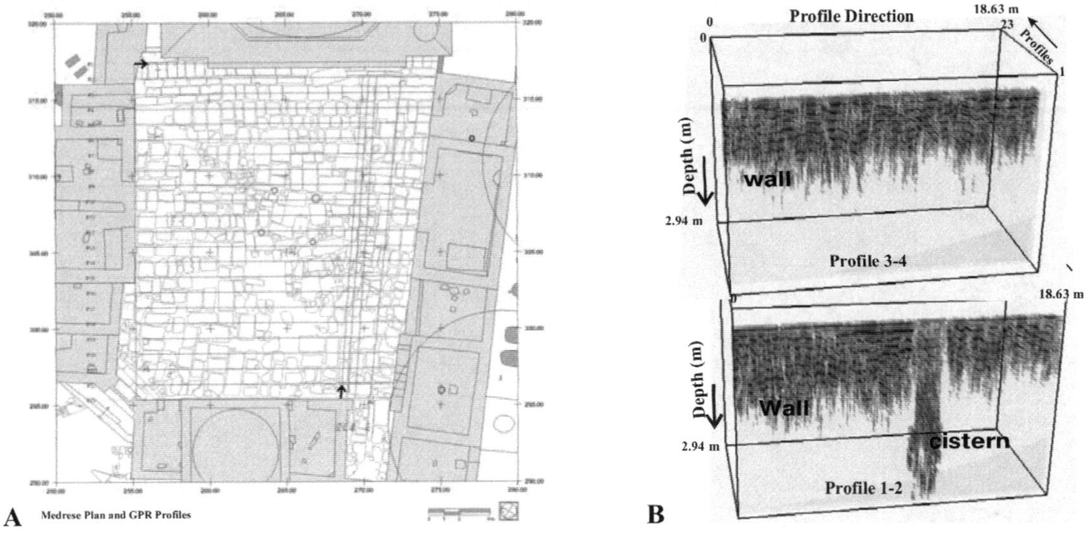

Figure 7. (A) Medresah plan and measured GPR profiles (P1 indicates profile 1), **(B)** Profile pillows of the profiles 1-2 and profiles 3-4 indicating wall structures and cistern structure.

4. Conclusions

The buried archaeological remains of the study area of the Ilyas Bey Mosque and its around were determined by GPR method successfully. We could show some results of them. We used new approximation to determine buried objects by using simplified amplitude-colour range and appropriate opacity function and obtained transparent imaging of the subsurface.

The petrographical of the building stone with their mortar and plaster are investigated under polarized microscope. The thin section results of the collected samples from the study area reveal that the building stones are mainly composed of marble, schist, limestone, sandstone and metagranite. Some monzonite and granite rocks block are also found in the complex of Ilyas Bey. The thin section studies of the mortars and plasters of the walls show that the aggregate compositions are mostly identical with the local rock formation of the south-western part of Aegean region.

The geological and mineralogical, petrographical studies reveal that the main building rock units of Miletus and Ilyas Bey Mosque were obtained from Meandros Massive in Western Anatolia. The granodiorite columns might be obtained from Kozak Pluton in the north of Bergama-Izmir direction of western Anatolia.

Acknowledgments

The study of 3D visualisation of GPR data was partially supported by Ankara University under projects BAP 2005-07-45-030. The authors would like to thank Ankara University, for supporting the projects. The authors also would like to thanks KA-BA architecture LTD. for field work organisation.

5. References

[1] MTA 2002 Regional geological map of Izmir 1/500.000. Mineral Research Directorate

[2] Kadıoğlu S *2008 J. Appl. Geophys.* **64** 109–14.

[3] Kadıoğlu S, Ulugergerli E U and Daniels J J 2006. 3D visualization to map cavities by GPR method: Dalaman Akkopru dam reservoir area, Mugla, Southwest Turkey. *11th International Conference on Ground Penetrating Radar(GPR2006), June 19-22, 2006 (Columbus-Ohio, USA)* CD of GPR 2006

[4] Koralay T, Kadıoğlu S and Kadıoğlu Y K 2007. *Environ Geol.* **52** 1387–97

[5] Daniels J J 2002 Updated Gphyzgpr IDL 5.5 for GPR interpretation. http://www.geology.ohio-state.edu/~jeff/

IOP Conf. Series: Earth and Environmental Science **2** (2008) 012022 doi:10.1088/1755-1307/2/1/012022

Native of the marble in ancient city, Nysa on the Meander of Hellenistic and Roman Period, Aydin- Western Anatolia - Turkey

Musa Kadıoğlu[1,3] and Yusuf Kagan Kadıoğlu[2]
[1]Ankara University Faculty of Letters Archeological Department Ankara, Turkey
[2]Ankara University Engineering Faculty Geol. Eng. Dept. Ankara, Turkey

E-mail: kadioglu@humanity.ankara.edu.tr, kadi@eng.ankara.edu.tr

Abstract. Nysa, one of the most important cities of Caria in Hellenistic and Roman period, is located on the highway connecting Aydin (ancient name Tralleis) and Denizli, at about three kilometres northwest of Sultanhisar in western Anatolia of Turkey. The archaeological remains of Nysa are located on the slopes on the side of the stream called Tekkecik. The buildings, streets and public squares of the ancient city were supported by vaulted substructures adapted to the topographic conditions. As to the foundation of the city, Strabo relates that three brothers named Athymbros, Athymbrados, and Hydrelos came from Lakedaimon to Caria, and founded there three cities named after themselves. The small rock samples from the building of theatre, stadium, basement of agora and tomb were collected and determined under polarized microscope and confocal Raman spectroscopy to find out the native of these rock sources. The results of these studies reveal that the main rocks of these structures are composed from white colour marble. These marbles have granoblastic texture and are composed of mainly pressure twinned calcite as coarse grain size. The confocal Raman specroscopical studies of reveal that the marble building stone of Nysa city are mainly obtained from Jurassic Cretaceous of Western Anatolia marble.

1. Introduction

Nysa, is an ancient city and is represented one of the most important archaeological settlement on the Meander. The ruins of the ancient city are now on the slope of Mossogis (today Aydın Dağları) mountain by Sultanhisar which is a district of Aydın Province of Turkey, 70 km east of Ephesus, the capital city of Ionian (figure 1).

The archaeological remains of Nysa are located on the slopes on the side of the stream called Tekkecik. The buildings, streets and public squares of the ancient city were supported by vaulted substructures adapted to the topographic conditions. The known parts of the city today include several public buildings: The most important surviving buildings include the agora, gerontikon, and the Roman public baths in the east, and in the west the gymnasium, stadium, library, in antis temple, nymphaeum, theatre, bridges and the Byzantine churches. The remains of the Hellenistic walls have disappeared.

[3] To whom any correspondence should be addressed.

The object of this study is to determine the small collected samples from the Nysa settlement through confocal Raman spectrometry and polarizing microscope to identify the nature of the building stone of this ancient city on the Meander.

2. History

As to the foundation of the city, Strabo relates that three brothers named Athymbros, Athymbrados, and Hydrelos came from Lakedaimon to Caria, and founded there three cities named after themselves [1]. Nysa was founded in the 3rd century BC by Antiochus I, son of the Syrian king Seleukos, on the site of an ancient settlement called Athymbra. The name Nysa is thought to come from a member of Seleukos' family.

Nysa developed greatly during the period of Roman dominance in Anatolia, both architecturally and culturally. It is one of the most important cities of Roman Asia Minor because of the density of its buildings and different cultural periods that it preserves. Strabo, the famous Roman geographer (63 B.C. – 21 A.D.), was educated in Nysa.

Present researches have so far brought no evidence of the existence of the city before the third century B.C. Extant architectural remains and inscriptions as well as small finds prove that the city was inhabited uninterruptedly from Hellenistic to Byzantine times.

The first excavations in Nysa were done by the German railroad engineer Walter von Diest between 1907 and 1909 [2]. After this initial research, further excavation was directed by the Greek archaeologist K. Kourounitios in the 1920s. An interdisciplinary team under the heel of Vedat İdil from Ankara University has been studying the ancient city since 1990 [3].

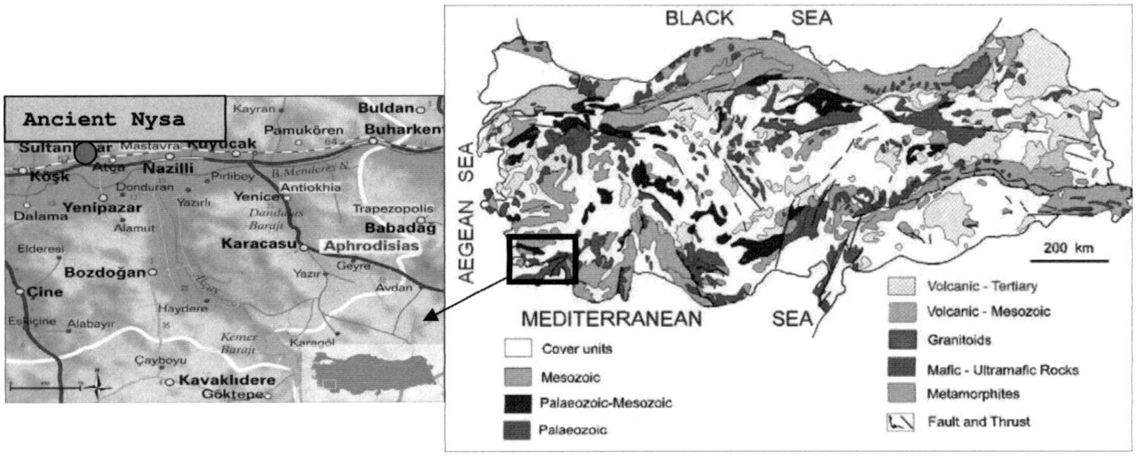

Figure 1. Location and simplified geological map of the region.

3. Building

3.1. Theater

The 2 km long, paved access road of Sultanhisar ends on a square in front of the theatre, which can still host about 10,000 spectators. It is located between the tunnel, the water reservoir in the north-south direction and the two bridges. This part of the city, offers the best topographic and climatic conditions for the construction of a Greek theatre, where in this case, the natural terrain had been used for the construction of the Cavea [4].

The whole complex (whose basic dimensions about 73 x 99 m) consists of the audience area (cavea), the orchestra, the proscenium and the stage building with a three-storey scaenae frons. The scaenae frons is completely built with marble opposing the cavea whereas the wall behind is masonary. During Roman renovations, the theatre preserved Greek features besides the newly attained

Roman character. We are dealing with an Anatolian-Roman building here. The general plan with its horseshoe-shaped cavea, a high proscenium, and a straight scaenae frons is typical of the Anatolian-Roman theatre. The scaenae frons plan with a straight wall surface and five doors enlivened by tabernacles is a characteristic feature of the Anatolian type (figure 2).

The examination and stylistic analysis of the ornamentation has revealed three construction phases for the theatre: The first phase of the theatre falls into the late Hellenistic and early Augustean period, i.e. the second half of the first century BC, based on profile series and ornamentation of architectural elements as well as epigraphic evidence. This dating is furthermore supported by Strabo's (64/63 B.C-20 AD) literary mention. This phase is connected to the enlargement of the entire complex including the cavea and the stage building.

Figure 2. View of the ancient city, Nysa on the Meander.

Two other construction phases followed in the second century AD, and applied only to the stage building, i.e. the scaenae frons and the proscenium façade. The second construction phase of the, then presumably two-storey, façade can be dated to the time of Hadrian, to the 20s and 30s of the second century AD, simply on the basis of the ornamented architectural pieces. The third construction phase of the theatre can be dated to the late Antonine and early Severian period, i.e. 180-200 AD, based on a stylistic comparison of architectural ornamentation. The fact that the architectural pieces were adapted from the Hadrianic period to their second use and mostly newly made with an ornamentation that corresponds to the third construction phase in the late Antonine and early Severian period, clearly shows that the Hadrianic façade must have been damaged or have collapsed for some reason. Presumably, the earthquake in 178 AD played a major role in that, although there is no epigraphic evidence from Nysa on the Meander to prove this. This dating is based only on architectural ornamentation, since there is neither a building inscription nor any other inscription from that time found in the theatre.

Most probably, the decoration school of Aphrodisias was responsible for the edification of the façade of the theatre in Nysa. Evidence of this is the identically worked motives on the one hand, and the structure of the stone used for the blocks on the other. The statues and sculptural works found in the theatre show a strong similarity of execution to the works in Aphrodisias.

On the basis of inscriptions dating back to Late Antiquity and small finds, one can say with certainty that the theatre remained in use at least until the sixth or seventh century AD, though its function in Late Antiquity is unknown. In the Byzantine period, no big construction activity seems to have occurred in the theatre.

3.2. Gerontikon

The Gerontikon is excellently preserved in Nysa on the Meander. Its cavea completely excavated and has twelve rows of seats and five stairways. It is located northwest of the marketplace (agora). The building has a rectangular plan with a shape as a large semicircle and had a width of 27.45 and a depth of 23.50 m (figure 2).

The council building, called the Gerontikon by Strabo, is certainly one of the best preserved structures of Nysa. It is understood that the Gerontikon, which comes from the word Gerousia meaning council of elders, is a bouleuterion, a council hal [5].

However, this important structure has not been examined completely, nor has it been properly introduced to the scientific world. W. von Diest made a plan of the structure during his excavations in 1907 and 1909, but this was only a general plan and many aspects of the structure were not examined in detail. Between 1921 and 1922 excavated K. Kourountiotis some parts of the Gerontikon's cavea, orchestra, scaenae frons and the stoa to the south. In these excavations, statue bases belonging to a rich man from Nysa, Sextus Julius Maior Antoninus Pythodoros, and to members of the Antonine imperial family, were found, along with many marble architectural blocks related to the façade of the structure. Considering the architectural decoration, the structural characteristics as well as the statutes, the remains of the Gerontikon could be dated to the middle of the second Century AD [6, 7, 8, 9].

3.3. Agora

The Agora (Market place) is located to the east of the Gerontikon and is essentially an open market area, measuring 113.5 x 130 meters [10]. Its interior dimensions are 88.5 x 102 m. It was surrounded by colonnades or stoas on all four sides. On the east and north sides were double colonnades of Ionic order while on the south and west sides stood single colonnades of Doric order. The main entrance, which is the connection adjacent to the west of Gerontikon, was probably in the middle of the western columned hall (figure 2). The differences in design and in the style of ornaments, makes it clear that the Agora was built in the 1st century B.C [3, 4].

3.4. Library

The library is one of the best preserved examples of an ancient library in Asia Minor [11]. This two storey building is accessible from the south. Square pilasters of buttresses project inwards from the west and east walls, with 2.50 meter deep alcoves between them. These alcoves which could also be observed in the Library of Celsus at Ephesus were designed to ensure shelves for storing bound volumes and papyrus rolls to be kept free from damp.

The east and west walls of the rectangular central area are still two storey (about 9 meters) high. Both of the walls have three large wall niches in each floor for wood cabinets. The dimensions of the building are 24.50 x 14.10 m. The room must have been made by a clay bricks of flat wooden roof.

In analogy to Trajanic Celsus-Library of Ephesus, whose hall is admittedly much larger (16.72 x 10.92 m), the Library of Nysa must have three monumental entrances on its south side, one of which is partially preserved; a marble door reveal decorated with clearly Hadrian ornamentation. The building must have been used about 120 and 140 AD.

3.5. Streets

The streets of Nysa, ancient city most probably were planned during the Hellenistic period and all the main and small streets on the direction of North-South and East-West parallel and perpendiculars to each other [12, 13]. The basements of all the rocks of these streets are composed of conglomerates. The compositions of these conglomerates are mostly pebbles of schist, marble, gneiss and fragments of granite within the carbonated cements (figure 2). The compositions of these conglomerates are most identical to the compositions of Mossogis of Aydin Mountain to the North of ancient Nysa city.

4. Petrography and Confocal Raman Spectrometry

Many small rocks, mortar and plaster samples were collected from ancient Nysa city settlement for the determination under the microscope and through confocal Raman spectrometer. The samples are prepared and polished for the analyses under the polarized microscopy and the Raman confocal micro-spectroscopy. The rock samples of Nysa city are mainly composed of marble, limestone (figure 3), conglomerate, mica schist and biotite gneiss and sandstone. The mortar of the Nysa building Stone mainly composed of aggregate and mortar. The average aggregates is 74% and average of the mortar is 26%. The aggregate of the mortar is mainly composed of marble, schist, limestone and gneiss. The matrix of theses mortar is composed of lime with fewer amounts of marble fragments

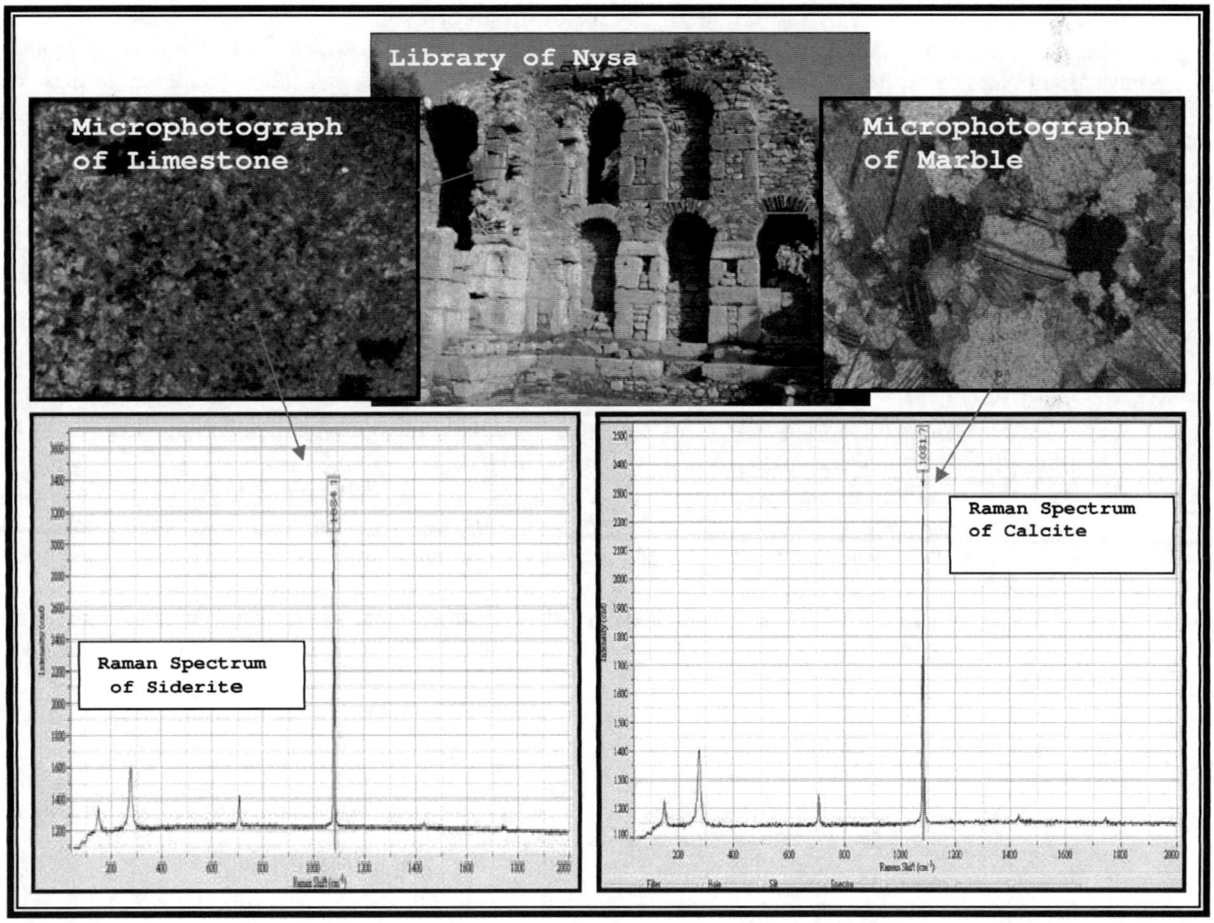

Figure 3. Microphotographs and Raman spectrums of the collected samples from the library of Nysa.

The samples were studied by a Confocal Raman (HR-800 HORIBA-JobinYvon, France) for identification of minerals. The confocal Raman spectrometer (CRS) is a well-known method for the analyses of minerals and however; CRS has not yet been widely applied on the archaeological mineral identification and determination. Raman measurements were performed with a Horiba Jobin-Yvon equipped with a laser operating at a wavelength of 633 nm in Ankara University Geological Engineering Department petrographical research laboartuary. An electrical cooled charge coupled device (CCD) detector was employed to acquire spectra, and the laser spot was focused on mineral surface with 10X, 50X or 100X long focused objectives, which allow a 65 and 13 mm working distance, respectively, and a lateral resolution of 5 and 2 μm respectively. Polarization of the incident laser beam was selected parallel to the preferential domain orientation of samples (y) and spectra were collected in a strict backscattering geometry. The intensity values of phonon modes for mineral were obtained using a Labspec 4.02 (Horiba/Jobin-Yvon) package, according to the Gaussian-Lorentzian mixed functions after subtracting a baseline and applying the initial approximations of modes position. The marble, limestone and some of the schist of the building stone are composed of calcite and some of the cracks of these rock units are filled with recrystallised calcite and siderite (figure 3).

5. Results

Nysa building stones are mainly composed of blocks of marble, limestone, metalimestone, metasandstone, schist, gneiss and conglomerate. The small rock samples from the building of theatre, stadium, basement of agora and tomb were collected and determined under polarized microscope and confocal Raman spectroscopy to find out the native of these rock sources. The results of these studies reveal that the main rocks of these buildings are composed from white colour marble. These marbles have granoblastic texture and are composed of mainly pressure twinned calcite as coarse grain size under the microscope. The confocal Raman spectroscopical studies reveal that the marble building stones of Nysa city are mainly obtained from Jurassic- Cretaceous carbonate rocks of Western Anatolia marble quarry in the vicinity region.

Acknowledgements

The authors would like to thank TUBITAK for the supporting the project, project Nr. 107K523.

6. References

[1] Strabo Geographica, 14. I. 46 (C 650)
[2] Diest W von 1913 *Nysa ad Maeandrum. Nach Forschungen und Aufnahmen in den Jahren 1907 und 1909* Jd I Ergh **10**, 1 ff
[3] İdil V 1999 Nysa ve Akharaka *Nysa and Acharaca* 1ff
[4] Kadıoğlu M 2006 Die scaenae frons des Theaters von Nysa am Maeander. Forschungen in Nysa am Maeander I, 1 ff *Roman Theatres. An Architectural Study* ed F Sear F pp 345–6
[5] Balty Ch 1991 Cvria Ordinis, recherches darchitecture et d'urbanisme antiques sur les curies provinciales du monde romain p 451
[6] Meinel R 1980 *Das Odeion* Untersuchungen an überdachten antiken Theatergebäuden p 319
[7] Gneisz D 1990 *Das antike Rathaus* Das griechische Bouleuterion und die frührömische Curia **40** 338–9
[8] Kadıoğlu M 1999 "Der opus sectile Boden aus dem Gerontikon-Buleuterion von Nysa ad Maeandrum", Studien zum Antiken Kleinasien IV, AMS **34** 175
[9] Kadıoğlu M 2008 "Ein Porträitfragment des Lucius Verus? aus dem Gerontikon von Nysa am Maander" *Vom Euphrat an den Bosporus. Kleinasien in der Antike. Festschrift für Elmar Schwertheim zum 65. Geburtstag. 2 Bde.* ed E Winter Asia Minor Studien **65** 359–66
[10] İdil V 2006 Neueste Grabungsergebnisse von der Agora in Nysa am Mäander. *Die Griechische Agora. Bericht über ein Kolloquium am 16. März 2003* W eds Hoepfner and L Lehmann (Germany: Berlin) pp 65–76
[11] Hiesel G and Strocka V M 2006 "Die Bibliothek von Nysa am Mäander: Vorläufiger Bericht über die Kampagnen 2002-2006" AA 2 Halbband pp 81–97

[12] İdil V and Kadıoğlu M 2007 "2005 Yılı Nysa Kazı ve Restorasyon Çalışmaları" KST 28.2 (2008) pp. 657–62

[13] İdil V and Kadıoğlu M 2008 "2006 Yılı Nysa Kazı ve Restorasyon Çalışmaları" KST 29.2 2007 (2008) pp. 215–18

Paleoseismic investigations along a key active fault within the Gulf of Corinth, Greece

I K Koukouvelas, S Kokkalas[1] and P Xypolias
Department of Geology, University of Patras, 26500 Patras, Greece

E-mail: skokalas@upatras.gr

Abstract. The study of paleoseismological and archaeological excavations provide clues for the evolution of Helike Fault, located along the westernmost end of the Gulf of Corinth, that displays high activity and exerts control on the landscape. In this study we present evidence from paleoseismic trenches which revealed well defined fault strands and clear colluvial stratigraphy. We focus on the two main segments of the Helike Fault and their implications on strong earthquake activity. The Helike Fault is a major tectonic structure that influenced the evolution of ancient settlements on the Helike Delta, from the Early Bronze Age through the Byzantine period, till present times. The eastern fault segment appears to control the southern Gulf morphology, while the western segment is controlling the large Aigion basin. Interbedded organic-rich soils and gravels dominate in all trenches. Fault strands that control successive scarp-derived colluvial deposits were identified within the trenches and indicate the continuous seismic activity along the fault trace. Co-seismic offsets, open cracks filled with debris and liquefaction related deformation was also recognized. At least seven seismic events were identified inside the excavated trenches, during the last 10 ka. The estimated vertical throw along the fault segments, observed within the trenches, is on the order of 1meter per event. Based on dating of colluvial wedges we estimated the Holocene slip rate on the Helike Fault, which shows an increase from ~0.3 mm/yr to 2 mm/yr in the last 2 ka. We consider the derived slip rates to be minimum values due to the implication of erosional effects and sediment accumulation from the upthrown block. The Helike fault appears to play a crucial role both in subsidence of the Helike delta plain and in shifting Kerynites river course that runs between the two Helike fault segments. The Helike Fault activity and the clustering of surface rupturing events on the Helike fault seems to fit well with the subsidence of the Helike Delta plain and its change from marsh to lake or pod over the last 5 Ka.

1. Geological setting-The Gulf of Corinth

The 120-km long Gulf of Corinth offers a typical example of rapid extension in a continental region [1] (figure 1). Active normal faults strongly control the landscapes bordering the Gulf of Corinth leading to major changes of fluvial geomorphology [2], shorelines [3] and the surface of the Helike Delta [4]. In all these examples, archaeology compliments the data from geology in the interpretation of landscape evolution [4, 5].

Strong earthquakes in the area, are related to offsets on the order 1 m [i.e. 6]. However, only one earthquake during 1861 is known to have ruptured the surface within the western part of the Gulf of Corinth (figure 2). The oldest historically described earthquake that occurred in the Gulf of Corinth

[1] To whom any correspondence should be addressed.

region destroyed the cities of Helike and Boura in 373 B.C. but the faults hosting this historic event remains unknown [7, 8].

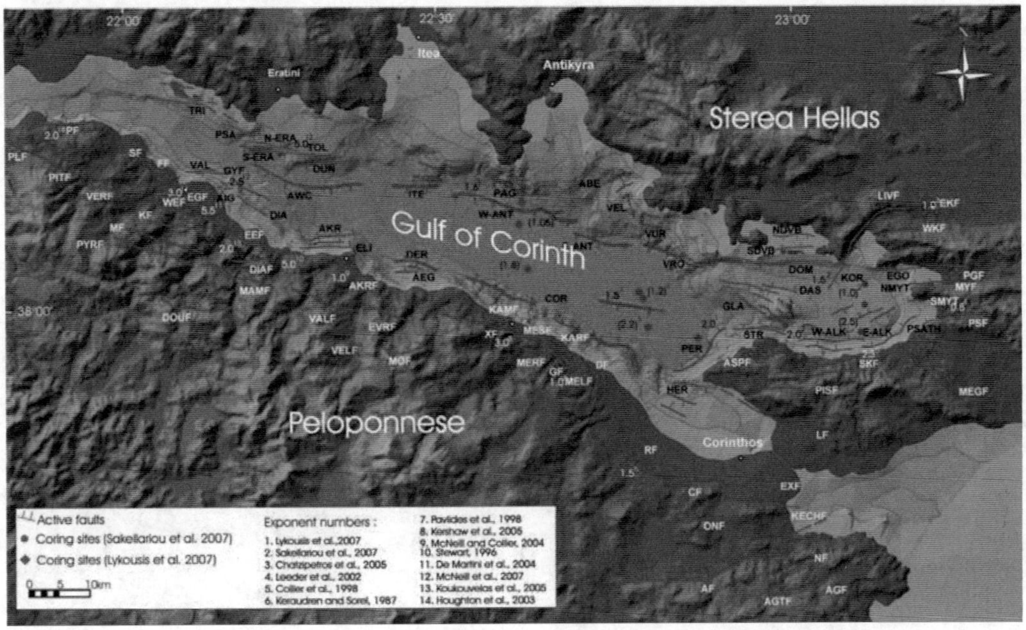

Figure 1. Tectonic map showing the active normal faults in the Gulf of Corinth and the inferred slip rates both for offshore and onshore faults (map from [9]).

2. The Helike active fault

The Helike Fault is a major tectonic structure controlling the landscape of the western Gulf of Corinth (figure 2). It strongly influenced the evolution of ancient settlements on the Helike Delta from the Early Bronze Age through the Byzantine period [4, 10].

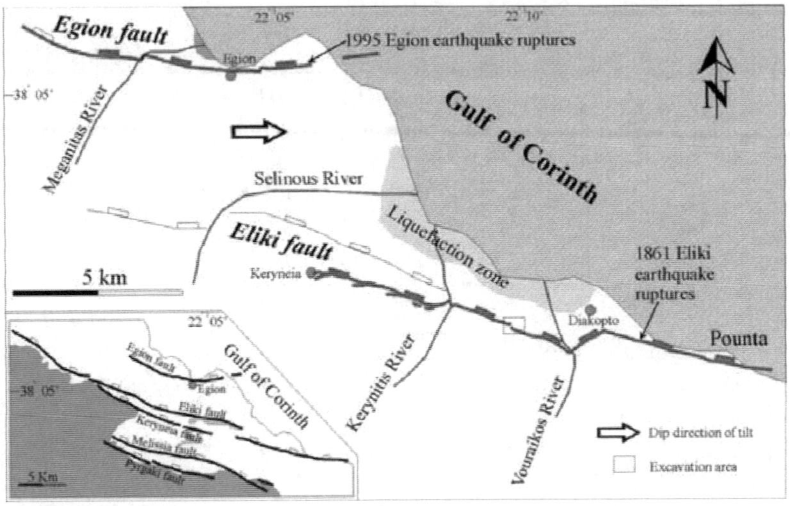

Figure 2. Structural map showing the Helike fault and the surface ruptures of the last strong 1861 earthquake event in the area.

The Helike Fault is divided into a western and eastern segment, 9 and 13 km long, respectively. The two fault segments dip north and their tips show an approaching configuration, forming the Helike step-over zone (figure 2). The easternmost segment appears to control the southern Gulf morphology. The western segment of the Helike Fault is also important in controlling the large Aigion basin [11].

2.1. Paleoseismological trenching and archaeological data

2.1.1. East Helike. Archaeology provides another mean to date the events within the paleoseismological trenches, as well as to identify the area most strongly affected by earthquakes (figure 3). Excavation of the trench T13 (figure 4) revealed the ruins of semi-circular structures built with stones of local psammite enclosing the position of a Late Roman tile-covered grave. Large clay tiles of Laconian type apparently belonging to tile-covered graves were previously discovered in trench Eli3 excavated east of T13 [12].

Figure 3. (Left) Early Bronze age architectural ruins showing offset due to earthquake. In this area archaeological evidence shows that buildings were thrown down by an earthquake accompanied by conflagration. Estimation of the event age is between 4500-4300 BP. (Right) Part of Hellenistic pebble mosaic floor covered by a thick destruction layer of fallen clay roof tiles and plaster [10].

Interbedded organic-rich soils and gravels dominate the T13 trench. The lowermost exposed unit is a brick-orange fluvial unit (BO_1 in figure 4, right). In the trench we recognized two fault strands, the "north" and "south" fault, which control colluvial wedges. In general, colluvial wedges related to the north fault are thinner and are buried by a well preserved colluvial sequence, which is controlled by the south fault (figure 4). The displacement along the north fault is as much as 2.7 m and controls the deposition of a set of three scarp colluvial units (labeled SC_1 ,$SC_2$and SC_3) and a structureless, silty-sand layer (L), as shown in Fig 4. Part of the SC_3 contains scattered pebbles, many of which have sub-vertically oriented long axes indicating liquefaction. The contact between the SC_3 colluvium and layer L is transitional. The L layer shows wavy contacts and sill-out within fluvial or colluvial gravels producing also an apparent blister-like bulge of SC_3. All these features indicate that layer L was produced by liquefaction. Since the lower part of the SC_3 also shows liquefaction effects, we infer that the liquefaction occurred some time after the deposition of the SC_3. The unit L is 1.5 m long and extends across the 5 m width of the trench, but its presumed dyke source is beyond the faces of the trench. The SC_3 colluvium covers the upper part of the scarp, attains a maximum thickness 0.7 m, including the part disturbed by liquefaction, and is overlain by 0.6 m of yellowish-white fluvial gravels.

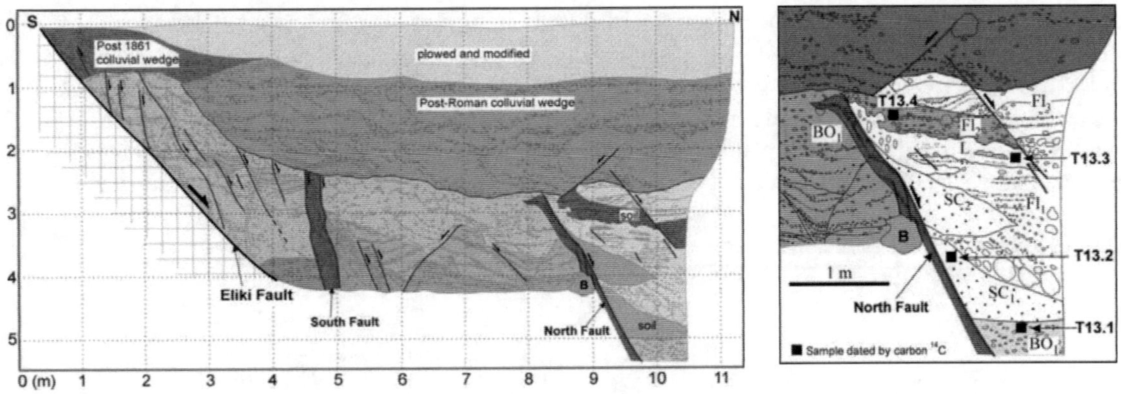

Figure 4. (Left) The paleoseismological trench T13 across the east Helike fault, showing the south and north fault strands. The post-Roman colluvial wedge covering much of the trench stratigraphy corresponds to SC4 colluvial wedge. (Right) Detailed view of the north fault strand with the successive colluvial wedges and the location of samples dated.

In another trench, excavated by the Helike Project just west of T13 in 2003, we discovered a well built tile-covered grave containing the skeleton of an adult man oriented E-W, not accompanied by grave goods [5]. Judging from the type of the tiles used for the construction of the graves, we dated them to the Late Roman period. The graves encountered in trenches T27, T13 and Eli3, were found, respectively, at 2.1, 2.5 and 3.0 m below the surface. They seem to have suffered destruction and erosion by the flow of a minor nearby torrent, which had formed a small scale fan underlying the graves. Above this fan is the colluvial deposit SC4 controlled by the south fault (figure 4). The discovery of the tombs allows one to distinguish between pre-Roman and post-Roman seismic events, and to infer that the third event on the north fault occurred between 1530 and 1230 BP.

2.1.2. West Helike. Across the West Helike Fault segment we excavated two trenches, given that promising sites for paleoseismological trenching are missing (figure 5).

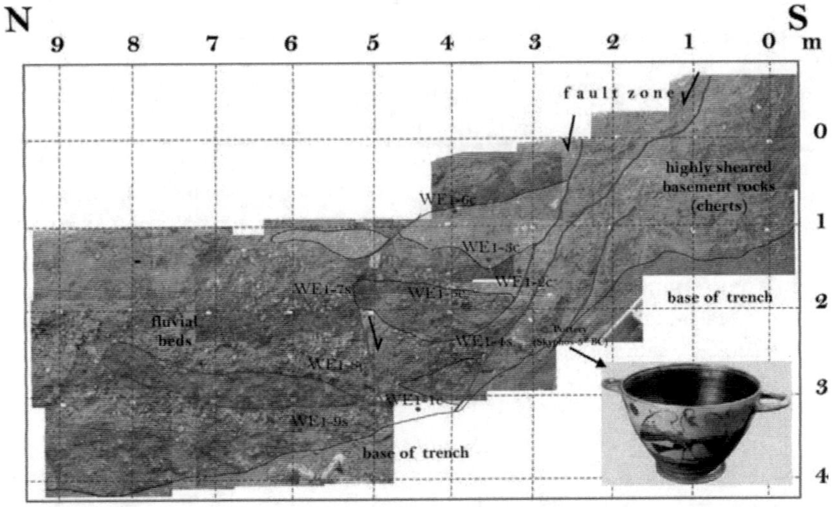

West Eliki Fault Section 1:east wall

Figure 5. Excavated paleoseismological trench across the west Helike fault, showing the succession of a series of colluvial wedges adjacent to the fault zone.

Within the trenches interbedded organic-rich soils and gravels are dominant. The West Helike 1 section is not unaffected by man-made disturbance and is characterized by a series of well preserved colluvial wedges within a scarp-graben system (figure 5). The graben is defined by an antithetic fault and includes a zone accumulating about 2.5 m of throw and tilt of the strata within the hanging-wall towards the Helike fault. Within this graben we recognized four colluvial wedges interpreted to be related with surface rupturing events. All events were dated from samples taken inside trenches with the C^{14} method. Two additional events were recognized within the We2 trench. Significant is that some of these events are directly comparable with events in the East Helike Fault segment and some of them are correlated with damages in the archaeological excavations (see also figure 6 right).

3. Results-Conclusions

1. The Holocene slip rate on both segments of the Helike Fault is not constant (figure 6, left). This result, together with the long-term average lip rate of 0.5 mm/yr for the Helike Fault estimated by [11] suggests that the fault built up its displacement during very short periods of high slip rate (>1.2 mm/yr) and much longer periods with low slip rate (<0.5 mm/yr).

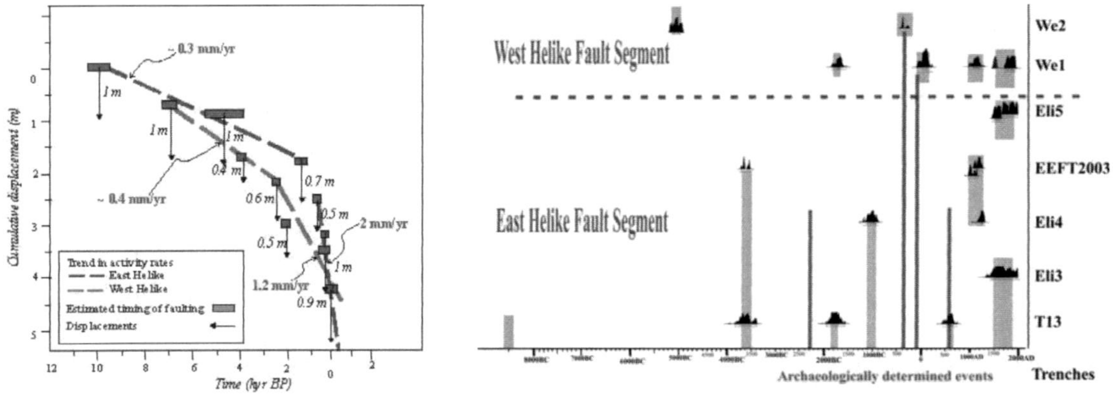

Figure 6. (Left) Summary of the displacement vs. time for the two segments of the Helike Fault. Estimated time for faulting and displacements for the East Helike Fault segment as is published in [12, 13], while data for the West Helike Fault segment as in this study. Dashed lines correspond to the slip rate of the Helike Fault. (Right) Calibration graph of all dated samples on the Helike fault trenches. The range of possible ages is shown only for the best level of confidence ($2\sigma = 95\%$). The results from the trenches show the age range for colluvial sedimentation.

2. The recognized switch from low to high slip periods is marked by surface rupturing earthquakes. Changes of the fault slip rate fit well with a model for clustering of earthquakes with periods of longer recurrence intervals. Thus, the Helike Fault activity resembles clustering similar to that of other major faults in the Eastern Mediterranean region, i.e. the North Anatolian Fault.
3. Recurrence intervals, estimated from our paleoseismological study, are almost identical for both faults in the order of 700-800 yrs, as well as the throw per event which is in the order of ~1m. However, in general the throw per event in the East Helike Fault segment is in general 20% larger than in the West Helike Fault.
4. For the last two events it appears that the two fault segments were activated as a linked fault system.

References

[1] Doutsos T and Kokkalas S 2001 *J. Struct. Geol.* **23** 455–42
[2] Koukouvelas 1998 *J. Geody.* **26** 502–13
[3] Stewart I 1996 *Annali di Geofisica* **39** 575–88
[4] Soter S and Katsonopoulou D 1999 *Geoarchaeology* **14** 531–63

[5] Katsonopoulou D 2003 *Tekmeria (Paestum)* **3** 205–16

[6] Roberts G P and Koukouvelas I K 1996 *Annali di Geophysica* **39** 619–46

[7] Marinatos S N 1960 *Archaeology* **113** 186–93

[8] Katsonopoulou D 2000 The Earthquake of 373 BC. Literary and Archaeological Evidence. *Ancient Helike and Aigialeia: Archaeological Sites in Geologically Active Regions Proceedings of the Third International Conference, October 2000* eds D Katsonopoulou, S Soter and I Koukouvelas (Athens, Greece: The Helike Society) p. 4.

[9] Zygouri V, Verroios S, Kokkalas S, Xypolias P and Koukouvelas I K 2008 *Tectonophysics* doi: 10.1016/j.tecto.2007.06.01

[10] Katsonopoulou D, Soter S and Koukouvelas I 2004 Historical earthquakes in the area of Helike. The evidence of geoarchaeology. *5th International Symposium on Eastern Mediterranean Geology, (Thessaloniki, Greece, 14–20 April 2004)* eds A A Chatzipetros and S B Pavlides (Thessaloniki, Greece: University of Thessaloniki) pp. 759–60

[11] Doutsos T and Poulimenos G 1992. *J. Struct. Geol.* **14** 689–99

[12] Koukouvelas I, Stamatopoulos L, Katsonopoulou D and Pavlides S 2001 *J. Struct. Geol.* **23** 531–43

[13] Koukouvelas I K, Katsonopoulou D, Soter S and Xypolias P 2005 *Terra Nova* **17** 158–64

Orogenic vs anorogenic lamproites in a single volcanic province: Mediterranean-type lamproites from Turkey

D Prelević[1,3], **C Akal**[2] **and S F Foley**[1]
[1] Institute of Geological Sciences, University of Mainz, Becherweg 21, D-55099 Mainz, Germany
[2] Dokuz Eylül Üniversitesi, Mühendisliği Bölümü, TR-35160 Buca, Izmir, Turkey

E-mail: prelevic@uni-mainz.de

Abstract. Lamproites are mantle-derived ultrapotassic volcanic rocks, derived from phlogopite-bearing harzburgitic source. The origin of the metasomatism which enriched their mantle source is either an old event with a metasomatic component derived from the convecting mantle, or is a more recent introduction of an already aged metasomatic component. Together with different trace element signatures, this view serves for a general distinction between anorogenic and orogenic lamproites worldwide. In Turkey, lamproitic volcanism resulted from interplay of subduction/collisional and postcollisional/extensional regime since Miocene until Pliocene, in the Western Anatolia-Aegean and Kırka-Afyon-Isparta region. We present new set of Ar-Ar age data together with geochemical data, including Sr, Nd and Pb isotopes, of all Turkish lamproites. Our study revealed the most intriguing geochemical data: although the majority of lamproites have an orogenic affinity, with radiogenic $^{87}Sr/^{86}Sr$, $^{207}Pb/^{204}Pb$, and unradiogenic $^{143}Nd/^{144}Nd$ and $^{206}Pb/^{204}Pb$, and high LILE/HFSE ratios, the lamproites from the most southern localities (Bucak area) exhibit geochemical features characteristic for anorogenic lamproites, with unradiogenic $^{87}Sr/^{86}Sr$, $^{207}Pb/^{204}Pb$, and radiogenic $^{143}Nd/^{144}Nd$ and $^{206}Pb/^{204}Pb$, coupled with rather smooth incompatible trace element pattern with low LILE/HFSE ratios and high concentrations of Nb and Ti. The common and coeval occurrence of anorogenic and orogenic lamproites in a single volcanic province suggests that geodynamic distinction of lamproites based only on geochemistry may be questioned.

1. Introduction

Since recognition of lamproites as important diamond carrier in the 1970s, their origin repeatedly attracted many researchers. They are mantle-derived volcanics characterized by severe enrichment in incompatible trace-elements combined with variable and extreme radiogenic isotope compositions. General agreement exists that a normal four-phase peridotitic mantle cannot represent their source, because a K-rich hydrous mineral, in most models phlogopite, is required. The origin of the metasomatism which enriched their mantle source is, however, a confusing issue, as lamproites typically have a very "ancient" isotopic signature which cannot always be related to known tectonic events: it is either an ancient event with a metasomatic component derived from the convecting mantle, or is a more recent introduction of an already aged metasomatic component. This view,

[3] To whom any correspondence should be addressed.

together with different trace element signatures, serves ultimately for a general distinction between anorogenic and orogenic lamproites (figure 1).

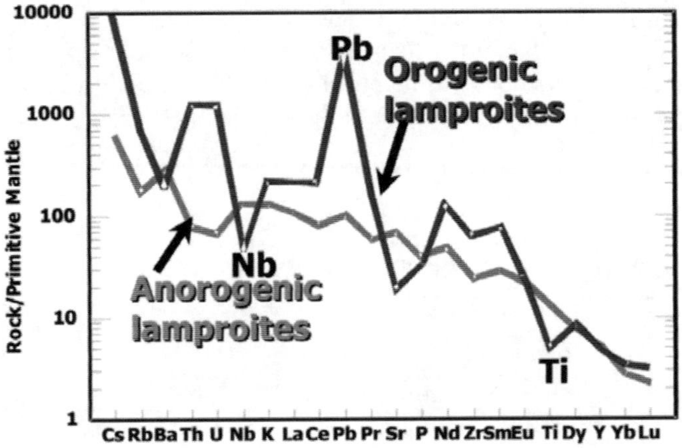

Figure 1. Orogenic vs anorogenic lamproites: trace element signature.

2. Mediterranean Tertiary volcanism

Lamproitic volcanism is widespread in the Mediterranean area (figure 2) ([1] and references therein). It presents an integral part of postcollisional volcanism within the Alpine-Himalayan belt, which took place between the late Eocene-Oligocene and the present day. Given their primitive character, lamproites are very often interpreted as potential candidates for parental melts of more voluminous postcollisional volcanics like trachybasalts, shoshonites and leucitites [2]. Coevally with postcollisional volcanism, Circum-Mediterranean anorogenic Na-alkaline Cenozoic igneous volcanism [3] apparently of sublithospheric mantle origin occurs all over Mediterranean. Na-rich alkaline volcanism has been reported within the Iberian Hercynian Massif, Apennine Peninsula, Serbia and Turkey.

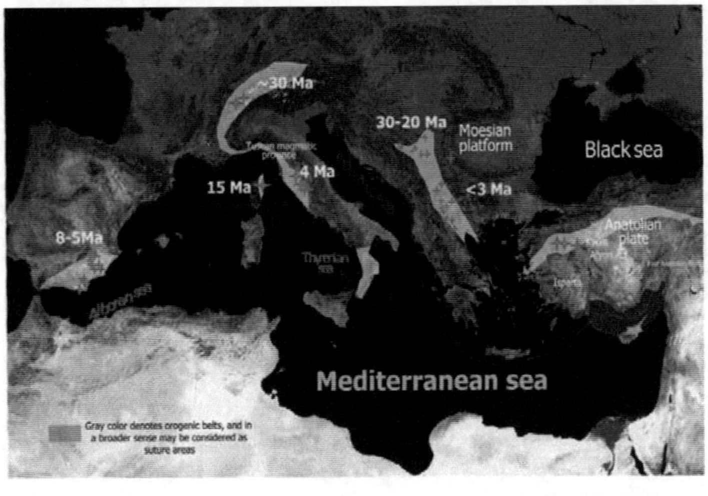

Figure 2. Distribution of Mediterranean lamproites with their ages.

3. Mediterranean lamproites

Most of the Mediterranean lamproites share common geochemical features with their worldwide orogenic counterparts (figure 3). Their isotopic signatures though, imply derivation from melts with three mantle components involved in their origin, characterized by contrasting geochemical features which appear in $^{206}Pb/^{204}Pb$, $^{87}Sr/^{86}Sr$ and $^{143}Nd/^{144}Nd$ space [4]: (1) a mantle source contaminated by crustal material, giving rise to crust-like trace element patterns and radiogenic isotope systematics, (2) an ultra-depleted mantle component derived from oceanic-type mantle accreted to older lithosphere during Mesozoic collisional processes [5], isotopically similar to Depleted (MORB) Mantle [6] and (3) a component originating from the convecting mantle. The presence of so many components demand multistage preconditioning of the lamproite mantle source, involving an episode of extreme depletion, followed by involvement of terrigenous sediments, and finally interaction with melts originating from the convecting mantle, some of which are probably carbonatitic [4].

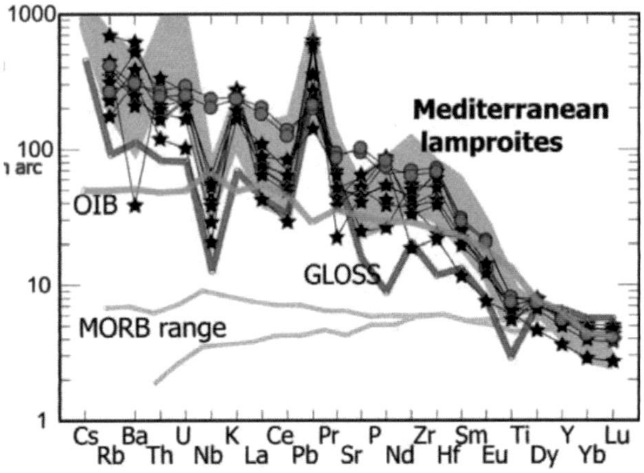

Figure 3. Trace element signature of Mediterranean lamproites. Black stars denote Turkish "orogenic", and red circles "anorogenic" lamproites.

4. Anorogenic lamproites in orogenic setting: Turkey

Although the majority of Mediterranean lamproites have an orogenic affinity, with radiogenic $^{87}Sr/^{86}Sr$, $^{207}Pb/^{204}Pb$, unradiogenic $^{143}Nd/^{144}Nd$ and $^{206}Pb/^{204}Pb$, and high LILE/HFSE ratios, the lamproites from few localities in Turkey exhibit geochemical features characteristic for anorogenic lamproites and generally resemble melts derived from convecting mantle. They have lower silica and higher CaO contents, with unradiogenic $^{87}Sr/^{86}Sr$, $^{207}Pb/^{204}Pb$, and radiogenic $^{143}Nd/^{144}Nd$ and $^{206}Pb/^{204}Pb$, coupled with rather smooth incompatible trace element pattern with low LILE/HFSE ratios, lower Pb peak and high concentrations of Nb and Ti (figures 3 and 4).

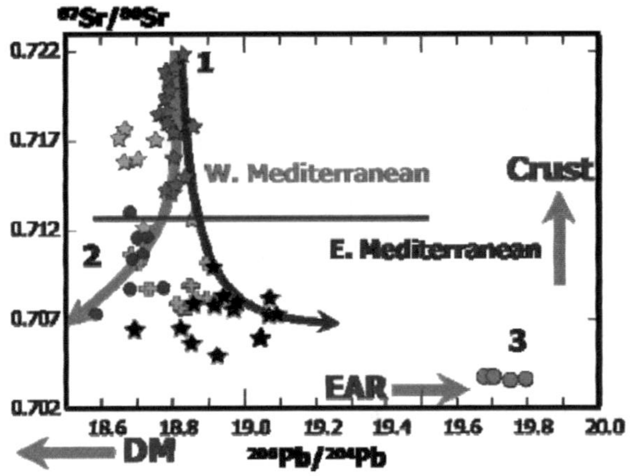

Figure 4. Pb vs Sr isotopes for Mediterranean lamproites: three mantle components.

Still, their K_2O contents is similar to that of anorogenic lamproites, and is near-constant (6.5 ± 2 wt.%) indicating buffering by residual phlogopite during melt segregation. This is because K is a stoichiometric component in phlogopite and the K_2O contents in the melt will be buffered as long as this phase is residual. The identification of residual phlogopite in the source of Mediterranean lamproites, including Turkish anorogenic ones, has important implications for the location of the magma sources. The phase relationships show that phlogopite is neither stable in the asthenospheric upper mantle nor in thermal plumes. In both cases temperature is the limiting parameter rather than pressure. Phlogopite is stable in the cold continental and oceanic lithospheric mantle.

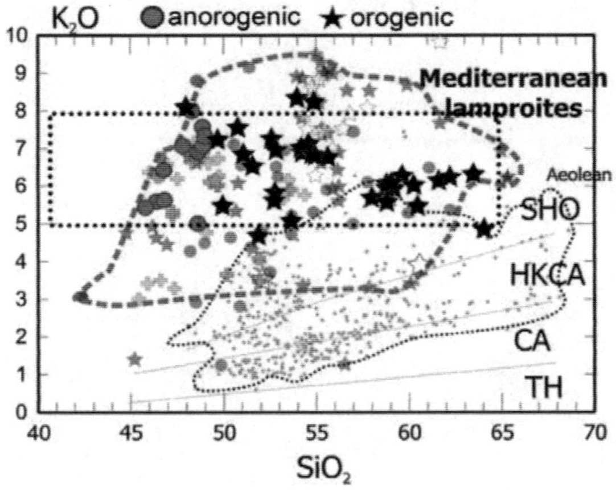

Figure 5. SiO_2 vs K_2O variation for Mediterranean lamproites. Note constantly high potassium contents in both "types" of Turkish lamproites.

5. Geodynamic scenario and petrogenetic model

General geodynamic environment during which Turkish volcanism has been triggered, is clearly postcollisional and extensional. It resulted after the collision of the Sakarya and Tauride continental blocks in the early Eocene e.g. [7]. It was proposed that during poscollisional orogenic collapse, the exhumation of several metamorphic core complexes has taken place. Lamproitic magmatism spatially surrounds the largest of them, the Menderes core complex (figure 6), and they temporally track all extensional episodes recorded in its exhumation (figure 7): the oldest lamproites occur in the 23–20 Ma coevally with the beginning of extensional deformation in the northern Menderes Core Complex; the major episodes of the tectonic exhumation of middle-crustal rocks in the central Menderes Massif is also the time of the major lamproitic volcanism in several locations surrounding Menderes; finally, the youngest southernmost lamproites occur contemporary with the last extensional episode recorded in Menderes as well, around 5 My, and correlates with initiation of N. Anatolian fault. The extensional episodes recorded within Menderes core complex most probably represent overall lithosphere response to the poscollisional isostatic requilibration and subsurface compensation.

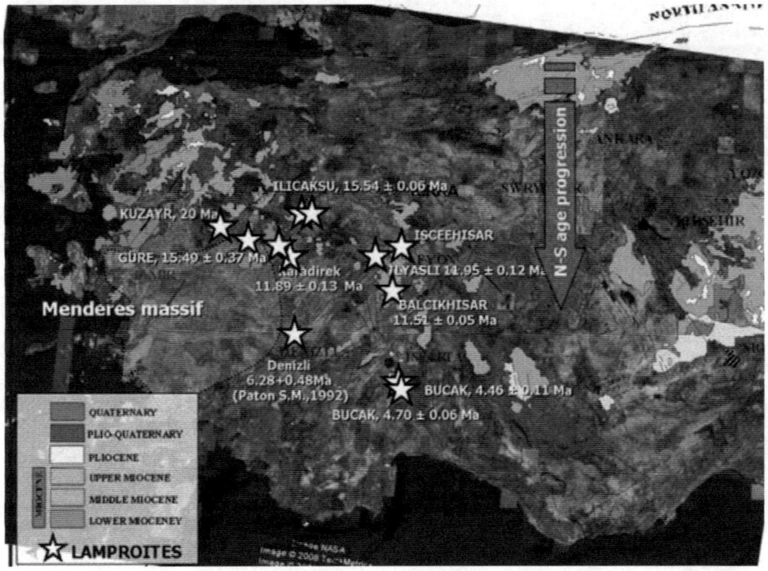

Figure 6. Time and space distribution of Turkish lamproites, together with other types of volcanism. The age of Kuzayr lamproites from [8]. The age of Denizli lamproites from [9].

When we put our geochemical data in the time-frame, we recognize a progressive southward younging of lamproitic volcanism coupled with an increasing intensity of asthenosphere-lithosphere interaction. Our geochemical together with age data imply that in Miocene (20 Ma), we have first lamproitic volcanism, which could be regarded as pure crustaly-metasomatized lithosphere-derived melts originated at the beginning of the lithosphere delamination.

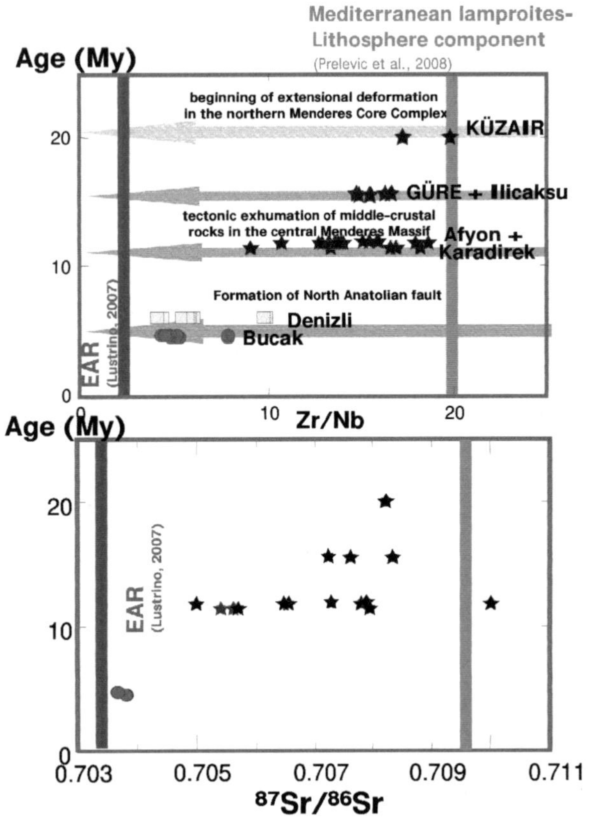

Figure 7. Zr/Nb and $^{87}Sr/^{86}Sr$ vs. age for Turkish lamproites illustrating pronounced lithosphere - asthenosphere interaction with time. Also three major extensional episodes recognized during the exhumation of Menderes core complex are denoted.

With age progression, we have advanced delamination and denudation of heterogeneously metasomatized lithospheric mantle which enabled more pronounced asthenosphere-lithosphere interaction, culminating in Pliocene, when "anorogenic" lamproites have erupted. According to our interpretation, whole lamproitic volcanism, originated by the melting of the lithospheric mantle, which underwent intense astenospherisation through time. We propose that lamproitic volcanism illustrate the reaction of the lithospheric mantle, including its delamination, with increase of intensity of lithosphere-asthenosphere interaction. Our petrogenetic model has implications for the interaction between sub-lithospheric mantle and the SCLM and melt generation in the SCLM.

6. References

[1] Prelević D, Foley S F and Cvetković V 2007 *Geological Society of America Special Paper* **418** 113–29
[2] Prelević D, Foley S F, Cvetković V and Romer R L 2004 *J. Petrol.* **45** 759–92
[3] Lustrino M andWilson M 2007, *Earth Sci. Rev.* **81** 1
[4] Prelević D, Foley S F, Romer R and Conticelli S 2008 *Geochim. Cosmochim. Acta* **72** 2125–56
[5] Prelević D and Foley S F 2007, *Earth Planet. Sci. Lett.* **256** 120–35
[6] Zindler A and Hart S 1986 *Annual Review of Earth and Planetary Sciences* **14** 493–571

[7] Okay A I, Satir M, Maluski H, Siyako M, Monie P, Metzger R and Akyüz S 1996 Paleo- and Neo-Tethyan events in northwest Turkey: Geological and geochronological constraints *Tectonics of Asia* eds A Yin and T M Harrison (Cambridge: Cambridge University Press) pp 420–41
[8] Ersoy Y and Helvaci C 2007 *Turkish Journal of Earth Sciences* **16** 117–39
[9] Paton S 1992 *Journal of Geological Society, London* **149** 1031–44

Cu isotope geochemistry of volcanogenic massive sulphide deposits of the eastern Pontides, Turkey

T B Housh[1,3] **and E Çiftçi**[2]
[1]Department of Geological Sciences, The University of Texas at Austin, 1 University Station C1100, Austin, TX 78712, USA
[2]Department of Geological Engineering, Niğde University, 51245 Niğde, Turkey

E-mail: housh@mail.utexas.edu

Abstract. A large number of volcanogenic massive sulfide (VMS) deposits are associated with Late Cretaceous to Eocene arc-like volcanic rocks in the eastern Pontides of NE Turkey. Cu isotope studies on thirteen VMS and two vein deposits were undertaken to examine the nature of copper isotope variations and to compare these with other VMS and black smoker deposits. δ^{65}Cu of chalcopyrite from these deposits range between +0.34 and -0.62‰. Chalcopyrite from the VMS deposits of the eastern Pontides have a mean δ^{65}Cu = -0.13‰. δ^{65}Cu of chalcopyrite is generally heavier than that of corresponding bornite. The range of δ^{65}Cu for chalcopyrite from VMS deposits in the eastern Pontides is larger than that observed from Alexandrinka, a Devonian VMS deposit in the southern Urals, but is significantly smaller than the up to 3‰ variations observed from individual modern sea-floor hydrothermal fields along modern mid-ocean ridges. The range of Cu isotope variation in VMS deposits from the eastern Pontides is interpreted to result from processes related to both oxidation and leaching of previously deposited copper by seawater and to its subsequent deposition elsewhere in the hydrothermal system.

1. Volcanogenic massive sulphide deposits in the eastern Pontides

The Pontide orogenic belt in northern Turkey extends from the Aegean Sea in the west to the eastern end of the Black Sea. The Pontides have a complex tectonic history resulting from the closure of a Paleotethyan oceanic basin in the Mesozoic and the accretion of the belt to Eurasia, and the subsequent closure of a Neotethyan oceanic basin culminating in the accretion of the Tauride-Anatolian platform to the Pontides during the mid-Eocene. Extension associated with the opening of the eastern Black Sea basin subsequently overprinted the eastern Pontides [1, 2].

Over sixty volcanogenic massive sulphide (VMS) deposits are associated with the Late Cretaceous to Eocene volcanic section of the eastern Pontides (between approximately 38° and 41°E) which represents the magmatic arc that developed above the northward-subducting Neotethyan oceanic plate. The volcanic sequence is up to 3 km thick and consists of a lower portion dominated by mafic and intermediate lavas while the upper portion of the section is dominated by felsic flows, breccias and pyroclastic deposits. The VMS deposits are associated with the felsic volcanic rocks of the upper part of the Late Cretaceous section [3, 4, 5].

[3] To whom any correspondence should be addressed.

The VMS deposits are massive, stratabound bodies which are commonly compared to the Kuroko deposits of Japan although many of the deposits in the eastern Pontides only exhibit some of the elements of a complete Kuroko deposit. A model of the VMS deposits in the eastern Pontides consists of an upper black ore zone with galena + sphalerite + chalcopyrite + bornite which is often overlain by either barite or gypsum. Beneath the black ore zone is a yellow ore zone consisting of pyrite + chalcopyrite + sphalerite, which in turn overlies a massive pyritic layer. The massive stratabound bodies are underlain by stringer ore in highly altered felsic volcanic rocks. The stringer ore consists of fracture-filling pyrite + chalcopyrite [3, 5, 6].

Pyrite occurs both early and throughout the paragenetic sequence. The beginning of pyrite deposition is followed successively by chalcopyrite(1), sphalerite, galena, tetrahedrite, tennantite, chalcopyrite(2), bornite and covellite, with the later minerals extensively replacing earlier deposited minerals [3, 5, 6]. Note that there is a transition from an early mineralogy where copper is present in a +1 valence state to a later mineralogy where copper is present in both +1 and +2 valence states.

Previous light stable isotope studies have demonstrated the importance of both magmatic and seawater components in the formation of these deposits. Massive, stringer and disseminated sulfides have $\delta^{34}S$ between 0 and 7‰, which is consistent with derivation of the sulfur from felsic igneous rocks in general. Barite and coarse-grained gypsum have $\delta^{34}S$ between 17.7 and 21.5‰, a few per mil higher than contemporaneous seawater sulfate, suggesting that perhaps these bodies were deposited in somewhat restricted basins [3, 7]. Quartz gangue has $\delta^{18}O$ between +8.8 and +11.8‰. This is within the range of fluids associated with felsic igneous rocks, but these values of $\delta^{18}O$ are also consistent with quartz-seawater fractionation between 200 and 325°C; temperatures which overlap those of fluid inclusions from these deposits [3].

2. Analytical methods

Samples of chalcopyrite and bornite for this study were dissolved in 10M HNO3. Aliquots of these solutions were diluted to between 500 and 1000 ppb Cu in 2% HNO3 and spiked with a solution of JMC Specpure Zn solution (Lot # 0-190W) to achieve a concentration of 500 ppb Zn. Chromatographic separation of Cu from the sulfide solutions was not deemed necessary given the evidence from the literature that Fe in the Cu solution does not result in measurable Cu isotopic variations, at least for the relatively simple copper sulfides studied [8, 9, 10].

Copper isotopes were measured on an IsoProbe magnetic-sector, multicollector ICP-MS in conjunction with an Aridus microconcentric desolvating nebulizer. The IsoProbe was operated with a forward power of 1350 Watts, 13.60 liters/minute cool gas, 1.20 liters/minutes intermediate gas, and 0.665 liters/minute nebulizer gas. The IsoProbe was run in hard-extract mode, and the hexapole was operated at 98% RF amplitude with Ar bled into the hexapole at a rate of 2.50 ml/minute. Masses ^{63}Cu, ^{65}Cu, ^{64}Zn, ^{66}Zn, ^{68}Zn, ^{62}Ni and ^{64}Ni were statically collected in faraday cups; the effect of isobaric overlap of ^{64}Ni on ^{64}Zn was found to be negligible.

The measured $^{65}Cu/^{63}Cu$ ratios were normalized for instrumental mass bias using $^{66}Zn/^{64}Zn$ through a phenomenological application of the generalized power law [11, 12, 13]. The primary standard was made from Alfa Aesar Puratronic copper foil (Lot # L02P23) and spiked with JMC Specpure Zn; fourteen analyses of this standard have $\delta^{65}Cu = -0.19 \pm 0.11$‰ relative to $^{65}Cu/^{63}Cu = 0.44560$ for the international standard ERM-AE467. Sixteen analyses of ERM-AE467 had a mean $\delta^{65}Cu = 0.00 \pm 0.08$‰.

3. Results

Cu isotope ratios were determined for chalcopyrite (cp) and bornite (bo) from the following thirteen VMS deposits: Cayeli (cp, bo), Harkoy (cp), Irsahan (bo), Kankoy (bo), Killik (cp), Koprubasi (cp), Kotarak Dere (cp), Kutlular (cp), Kutonit (cp), Kuvarshan (cp, bo), Lahanos (cp, bo), Murgul (cp, bo), and Yaylabasi (cp). Cu isotope ratios were also determined for chalcopyrite from the two following vein deposits: Ordu Bakacak (cp) and Oksuruk (cp). The results are summarized in table 1 and the histogram in figure 1.

Table 1. Summary of copper isotope variations for chalcopyrite and bornite from VMS and vein deposits in the eastern Pontides.

	δ^{65}Cu, ‰	N
VMS chalcopyrite	-0.13 ± 0.28[a]	12
VMS bornite (all)	-0.32 ± 0.36	6
Group A VMS bornite	-0.55± 0.16	4
Group B VMS bornite	+0.13 ± 0.23	2
Δ(chalcopyrite-bornite) (all)	+0.24 ± 0.38	4
Group A cp-bo pairs	+0.40 ± 0.25	3
Group B cp-bo pairs	-0.24	1
Vein-related chalcopyrite	+0.35 ± 0.14	2

[a]One standard deviation.

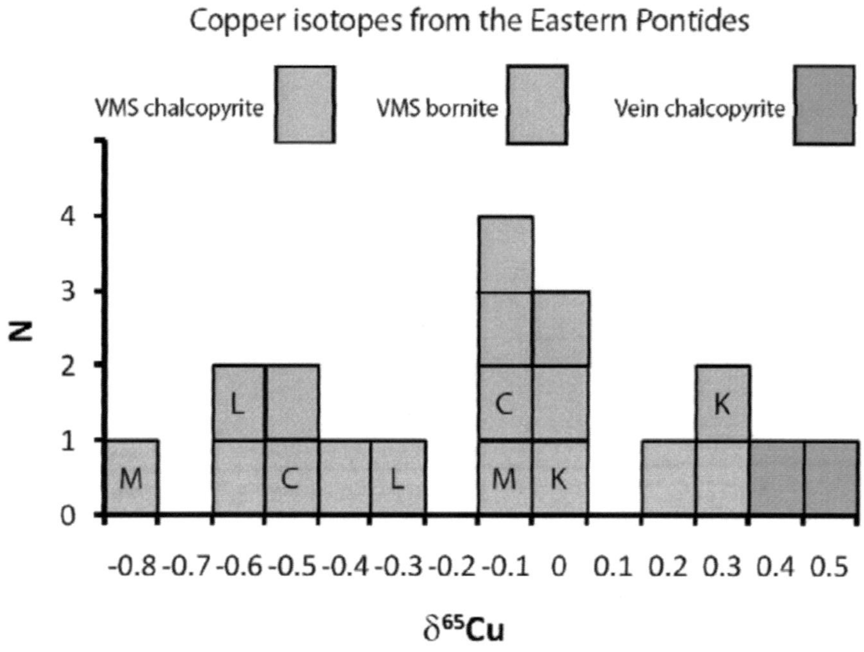

Figure 1. Histogram of copper isotope compositions for chalcopyrite and bornite from VMS and vein-related deposits of the eastern Pontides. Chalcopyrite – bornite pairs are related by the following letters: C – Cayeli, K – Kuvarshan, L – Lahanos, M – Murgul.

From the data presented in table 1 and figure 1 some basic observations can be made: 1) copper isotope compositions for chalcopyrite from VMS deposits cluster around 0‰. 2) Copper isotope compositions for chalcopyrite from vein-related deposits are invariably (although admittedly N is small) heavier than for chalcopyrite from the VMS deposits. 3) The bornite measurements can be divided into two groups; those with very light copper (Group A; lighter than all but one chalcopyrite measurement), and another with relatively heavy copper (Group B). The observation that the copper isotope composition of chalcopyrite from the VMS deposits cluster around 0 is consistent with the copper being of magmatic origin, although other processes must be responsible for the generation of the dispersion in these values. The heavier copper observed in chalcopyrite from the vein-related deposits suggests the copper in these deposits was recycled and were derived from materials in which copper had undergone some mechanism of fractionation before being deposited in the veins.

The copper isotope data for chalcopyrite from VMS deposits in the eastern Pontides are compared to published data of chalcopyrite from other VMS and black smoker deposits from modern hydrothermal vent fields along mid-ocean ridges in figure 2, below. It should be noted that while the range in δ^{65}Cu for chalcopyrite from VMS deposits in the eastern Pontides (+0.34 to -0.62‰) is somewhat larger than that observed at Alexandrekina, a Devonian VMS deposit in the southern Urals, it is still considerably lower than the up to 3‰ variation found in chalcopyrite from black smoker deposits in individual hydrothermal vent fields along modern mid-ocean ridges. This suggests that the processes responsible for the generation of copper isotope variation in arc-related volcanogenic massive sulphide deposits are either different than or less intense than those responsible for the generation of copper isotope variability in mid-ocean ridge-related volcanogenic massive sulphide deposits.

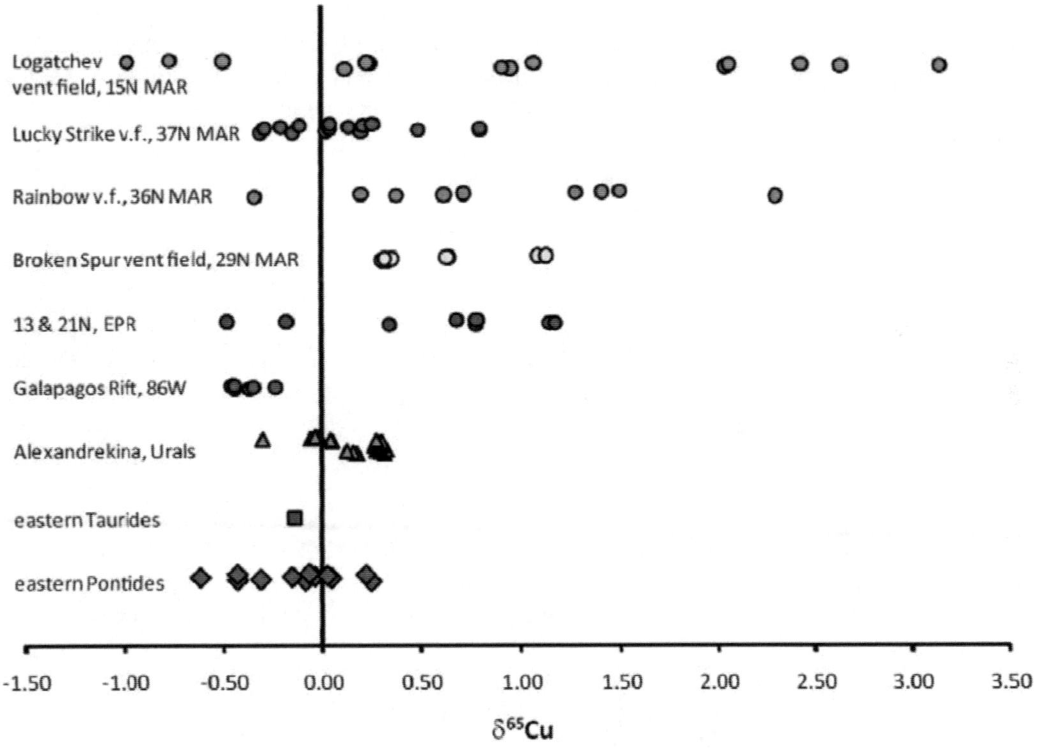

Figure 2. Copper isotope compositions of chalcopyrite from the volcanogenic massive sulphide deposits of the eastern Pontides as well as data from one deposit in the eastern Taurides, the Devonian Alexandrinka VMS deposits in the southern Urals [14], and for black smoker deposits from hydrothermal vent fields (v.f.) along mid-ocean ridges (MAR: Mid-Atlantic Ridge; EPR: East-Pacific Rise) [8, 15].

4. Copper isotope fractionation in VMS deposits from the eastern Pontides

The main processes responsible for the fractionation of copper isotopes at low temperature are believed to be: 1) redox reactions; e.g., oxidation of Cu(I) to Cu(II) can produce up to a 3 to 4‰ shift towards heavy copper [16]. And 2) biogenic processes; e.g. *Thiobacillus ferroxidans* incorporates isotopicaly heavy copper, producing fractionations as great as 5.6‰ [17]. Although diverse biotic communities are associated with black smokers, the lack of knowledge about their ability to fractionate isotopes of transition metals precludes invoking biogenic processes as a mechanism responsible for the generation of copper isotope variation in chalcopyrite from VMS deposits at this time. As shown below, however, a qualitative model relying solely upon redox processes is capable of generating the copper isotope variations observed in the VMS deposits of the eastern Pontides.

As noted above, ore studies have shown both the pervasiveness of replacement of earlier sulphides by later sulphides during the evolution of VMS deposits in the eastern Pontides, and that, furthermore, the oxidative state of copper in these deposits evolves from a mineralogy involving Cu(I) to one involving Cu(I) and Cu(II). On the basis of these observations as well as the results of the light stable isotopes discussed above we would like to suggest the following model for the generation of copper isotope variability in VMS deposits of the eastern Pontides. The first stage involves the percolation of seawater into a massive sulphide deposit containing copper initially with δ^{65}Cu 0‰; the seawater oxidizes and dissolves some of the primary copper minerals. This produces a significant shift toward heavier copper isotope compositions in solution. Mass balance requires that subsequent precipitation of chalcopyrite in a mineral assemblage containing phases that also incorporate Cu(II) from this solution will result in the precipitation of chalcopyrite heavier than 0‰. Overall, however, the massive sulphide pile is relatively reducing, and further percolation and cooling of the oxidized copper-bearing seawater through the pile will result in some degree of reduction of the copper in solution resulting in the generation of even lighter chalcopyrite precipitates. While copper isotope variations are generated by this process, the net effect will still be a VMS deposit with an average δ^{65}Cu = 0‰ as is observed for the deposits in the eastern Pontides.

5. Conclusions

The extent of copper isotope variation within VMS deposits from the eastern Pontides is similar to the extent of variation observed in the Devonian island-arc related Alexandrinka VMS deposit in the southern Urals. Like the VMS deposits at Alexandrinka, the copper isotope variation in the VMS deposits of the eastern Pontides is controlled by oxidation and reduction during seawater percolation through the sulfide pile. The extent of copper isotope variation in arc-related VMS of the eastern Pontides and the southern Urals is significantly restricted relative to the range of variation observed in modern ridge-related sulfide deposits: could this reflect weaker hydrothermal circulation associated with arc-related VMS deposits?

6. References

[1] Yilmaz Y, Tüysüz O, Yiğitbaş E, Genç Ş C and Şengör A M C 1997 Geology and tectonic evolution of the Pontides *Regional and Petroleum Geology of the Black Sea and Surrounding Region (AAPG Memoir 68)* ed A G Robinson (Tulsa: AAPG) chapter 11 pp 183–226

[2] Rice S P, Robertson A H F and Ustaömer T 2006 Late Cretaceous-Early Cenozoic tectonic evolution of the Eurasian active margin in the central and eastern Pontides, northern Turkey *Tectonic Development of the Eastern Mediteranean Region (Geol. Soc. Lond. Spec. Publ. vol 260)* ed A H F Robertson and D Mountrakis (London: The Geological Society) pp 413–45

[3] Çiftçi E 2000 *Mineralogy, Paragenetic Sequence, Geochemistry and Genesis of the Gold and Silver Bearing Upper Cretacous Mineral Deposits, Northeastern Turkey* (Rolla, Missouri: University of Missouri-Rolla, unplub. PhD dissertation) pp 278

[4] Özgür N 2003 Volcanogenic massive sulphide deposits (VMSD) in the East Pontic metallotect, NE Turkey *Mineral Exploration and Sustainable Development: Proceedings of the Seventh Biennial SGA Meeting on Mineral Exploration and Sustainable Development (Athens, Greece, 24-28 August 2003)* ed D Eliopoulos (Rotterdam: Millpress) pp 1217–20

[5] Çiftçi E, Yalcin G and Hagni R D 2004 Significant characteristics of volcanogenic massive sulphide deposits of the eastern Pontides (NE Turkey): a comparison to Japanese Kuroko deposits Applied Mineralogy: Developments in Science and Technology: Proceedings ICAM 2004 (Águas de Lindoia, Brazil, 19-24 September 2004) ed Pecchio M, Andrade F R D, D'Agostino L Z, Kahn H, Sant'Agostino, L M and Tassinari M M M L (São Paulo: ICAM-BR) pp 873–5

[6] Çiftçi E and Hagni R D 2005 *Türkiye Jeoloji Bülteni* **48** 55–64

[7] Çağatay M N and Eastoe C J 1995 *Mineral. Deposita* **30** 55–66

[8] Zhu X K, O'Nions R K, Guo Y, Belshaw N S and Rikard D 2000 *Chem. Geol.* **163** 139–49

[9] Graham S, Pearson N, Jackson S, Griffin W and O'Reilly SY 2004 *Chem. Geol.* **207** 147–69

[10] Larson P B, Maher K, Ramos F C, Chang Z, Gaspar M and Meinert L D 2003 *Chem. Geol.* **201** 337–50

[11] Albarède F, Telouk P, Blichert-Toft J, Boyet M, Agranier A and Nelson B 2004 *Geochim. Cosmochim. Acta* **68** 2725–46

[12] Maréchal C N, Télouk P and Albarède F 1999 *Chem. Geol.* **156** 251–73

[13] Wombacher F and Rehkämper M 2003 *J. Anal. Atom. Spect.* **18** 1371–75

[14] Rouxel O, Fouquet Y and Ludden J N 2004 *Econ. Geol.* **99** 585–600

[15] Mason T F D, Weiss D J, Chapman J B, Wilkinson J J, Tessalina S G, Spiro B, Horstwood M S A, Spratt J and Coles B J 2005 *Chem. Geol.* **221** 170–87

[16] Zhu X K, Guo Y, Williams R J P, O'Nions R K, Matthews A, Belshaw N S, Canters G W, de Waal E C, Weser U, Burgess B K and Salvato B 2002 *Earth and Planet. Sci. Lett.* **200** 47–62

[17] Mathur R, Ruiz J, Titley S, Liermann L, Buss H and Brantley S 2005 *Geochem. et Cosmochem. Acta* **69** 5233–46

Non-Andersonian conjugate strike-slip faults: Observations, theory, and tectonic implications

An Yin[1,3] **and Michael H Taylor**[2]
[1]Department of Earth and Space Sciences and Institute of Geophysics and Planetary Physics, University of California, Los Angeles, Los Angeles, CA 90025-1567, USA
[2]Department of Geology, University of Kansas, 1475 Jayhawk Blvd., Lawrence, KS 66044, USA

E-mail: yin@ess.ucla.edu

Abstract. Formation of conjugate strike-slip faults is commonly explained by the Anderson fault theory, which predicts a X-shaped conjugate fault pattern with an intersection angle of ~30 degrees between the maximum compressive stress and the faults. However, major conjugate faults in Cenozoic collisional orogens, such as the eastern Alps, western Mongolia, eastern Turkey, northern Iran, northeastern Afghanistan, and central Tibet, contradict the theory in that the conjugate faults exhibit a V-shaped geometry with intersection angles of 60-75 degrees, which is 30-45 degrees greater than that predicted by the Anderson fault theory. In Tibet and Mongolia, geologic observations can rule out bookshelf faulting, distributed deformation, and temporal changes in stress state as explanations for the abnormal fault patterns. Instead, the GPS-determined velocity field across the conjugate fault zones indicate that the fault formation may have been related to Hagen-Poiseuille flow in map view involving the upper crust and possibly the whole lithosphere based on upper mantle seismicity in southern Tibet and basaltic volcanism in Mongolia. Such flow is associated with two coeval and parallel shear zones having opposite shear sense; each shear zone produce a set of Riedel shears, respectively, and together the Riedel shears exhibit the observed non-Andersonian conjugate strike-slip fault pattern. We speculate that the Hagen-Poiseuille flow across the lithosphere that hosts the conjugate strike-slip zones was produced by basal shear traction related to asthenospheric flow, which moves parallel and away from the indented segment of the collisional fronts. The inferred asthenospheric flow pattern below the conjugate strike-slip fault zones is consistent with the magnitude and orientations of seismic anisotropy observed across the Tibetan and Mongolian conjugate fault zones, suggesting a strong coupling between lithospheric deformation and asthenospheric flow. The laterally moving asthenospheric flow may have been driven by the converging cratons with thick mantle lithosphere. This may have caused the shallow asthenosphere below a region sandwiched between the cratons to be squeezed out laterally.

1. Scope of work

Our ability to interpret lithospheric deformation depends critically on the knowledge that relates the observed fault geometry to the causative stress state. The most commonly used relationship in this regard is the Coulomb fracture criterion, which forms the basis for the famed Anderson fault classification (or "theory") of *X-shaped* conjugate faults at ~30° from the maximum compressive-

[3] To whom any correspondence should be addressed.

stress (σ_1) direction. [1, 2, 3] all noted that this pattern is not sustainable under finite-strain deformation. When examining conjugate fault systems at orogenic scales, one may also find that the X-shaped systems rarely occur in nature. Faults in dip-slip systems tend to have a single dip direction [see 4 for contractional systems and 5 and 6 for extensional systems]. Similarly, strike-slip conjugate systems tend to exhibit a *V-shaped* rather than X-shaped pattern. Not only do the V-shaped conjugate strike-slip faults defy the predicted X-shaped geometry, their orientations are also inconsistent with that inferred from the Coulomb fracture criterion in that they typically lie at 60-75° from the σ_1 direction. This type of conjugate strike-slip fault system occurs widely in the Alpine-Himalayan collisional system with prominent examples in the eastern Alps [7,8], Turkey [9, 10, 11], Afghanistan [12, 13], Tibet [14, 15, 16], Mongolia [e.g., 17, 18], Indochina [19], and Gulf of Thailand [20, 21, 22]. Similar structures also occur in subduction zones such as the Venezuela Andean conjugate faults that host large hydrocarbon traps [e.g., 23, 24].

Although V-shaped conjugate faults were long noted in the context of extrusion tectonics [25, 26], analogue and slip-line theory models all failed to reproduce the observed fault geometry [e.g., 7, 8, 26, 27, 28, 29]. Because V-shaped conjugate strike-slip faults are dominant features in collisional orogens and may have accommodated significant continental convergence [e.g., 19], it is imperative to understand the dynamic origin and kinematic evolution of these important yet poorly understood structures. In this paper we present a new hypothesis for the development of the V-shaped conjugate faults by advocating the important role of orogen-parallel asthenospheric flow. Our model has key implications for addressing two fundamental questions in the studies of continental dynamics: (1) do continents deform in a continuum or micro-plate fashion [30, 31]? and (2) is upper and lower crust coupled during continental collision [32; cf., 33]?

Several hypotheses were proposed to explain the non-Andersonian conjugate strike-slip faults. First, they could have developed by faults initiated at ~30° from the maximum principal stress (σ_1) direction following the Coulomb fracture criterion, reaching their current orientations by later vertical-axis rotation via bookshelf faulting or distributed deformation [2, 34, 35, 36]. Alternatively, the σ_1 and σ_3 directions may switch with time, causing the sense of fault slip to reverse, creating a non-Andersonian fault pattern. Finally, pre-existing anisotropy can also produce a non-Andersonian fault geometry [e.g., 37, 38, 39]. One may also consider applying the von Mises yield criterion to explain the observed conjugate faults. The criterion predicts two orthogonal sets of faults along the maximum strain-rate directions (i.e., slip lines), which are bisected by the principal stresses at 45°. Not only the predicted fault pattern still departs significantly from the observed conjugate faults oriented at 60-75° from the σ_1 direction, the prefect plasticity has never been observed experimentally for rock deformation [e.g., 40].

Experimentally, different faults can be produced under coaxial or non-coaxial strain conditions. The former results in conjugate faults as described by the Coulomb fracture criterion while the latter generates Riedel (R), conjugate Riedel (R') and primary (P) shears [41, 42]. More recently, careful experimental work shows that the formation of P and R' shears under non-coaxial deformation only develops in wet clays with exceedingly high cohesion, inappropriate for most rocks; for dry sand with no cohesive strength only R shears develop [43].

The different fault patterns under coaxial and non-coaxial conditions suggest that kinematics of deformation (e.g., velocity field) must also play a controlling role in fault formation. For coaxial deformation under a pure-shear condition, the velocity and strain fields are

$$u = Ax, \quad v = -Ay \tag{1}$$

$$\dot{\varepsilon}_{xx} = A, \quad \dot{\varepsilon}_{yy} = -A, \quad \dot{\varepsilon}_{xy} = 0 \tag{2}$$

and the corresponding vorticity describing local rigid-body rotation is

$$\omega_z = \frac{\partial v}{\partial x} - \frac{\partial u}{\partial y} = 0 \quad (3)$$

where u and v are velocity components, $\dot{\varepsilon}_{xx}$, $\dot{\varepsilon}_{yy}$, and $\dot{\varepsilon}_{xy}$ are strain rate components, and A is a constant. For non-coaxial deformation under a simple-shear condition, the velocity and strain-rate fields are

$$u = Ay, \ v = 0 \quad (4)$$

$$\dot{\varepsilon}_{xx} = \dot{\varepsilon}_{yy} = 0, \ \ \dot{\varepsilon}_{xy} = \frac{1}{2}A \quad (5)$$

and the vorticity (a negative sign for clockwise rotation in a right-handed reference frame) is not zero and is related to the velocity field by

$$\omega_z = \frac{\partial v}{\partial x} - \frac{\partial u}{\partial y} = -A \quad (6)$$

Using the above relationships, we can examine the relationship between the stress state and velocity field during fault formation. Under a pure-shear condition, the stress state generates two conjugate faults oriented at 30° from the σ_1 according to the Coulomb fracture criterion. As the vorticity is zero, the required rotation in the fault zones during their initiation is uninhibited by the associated velocity field. This is not the case, however, for simple-shear deformation in which σ_1 is oriented at an oblique angle from the shear boundary and should have the tendency to produce conjugate faults (R and R') with opposite sense of shear. The required rotation by the stress state for R shears is favored as it is promoted by the vorticity of the velocity field in the same direction. In contrast, formation of R' shears is not favored as the required rotation induced by the stress state is opposite to the direction of vorticity. The above analysis of combined stress state and deformation kinematics forms the basis for the formulation of a new hypothesis for the development of the V-shaped non-Andersonian conjugate strike-slip faults. Below, we use the relationship between fault geometry and GPS velocity field in Tibet to illustrate this concept.

Obtaining a complete velocity field at the time of fault formation is difficult, as most observable active faults of orogenic scales have life spans of at least a few million years, if not tens of millions of years [e.g., 44 for the Himalayan-Tibetan orogen]. To get around this issue, we may take the approach by first determining the current GPS velocity field across the youngest active conjugate strike-slip systems and then examining the temporal evolution of the faults to see if the past fault evolution is consistent with the current velocity field. Below, we use GPS data across the Central Tibet Conjugate Fault Zone to illustrate this approach. [45] converted the Tibetan GPS velocity field obtained by [46] into a strain-rate field, which shows that active deformation at the GPS time scale is consistent with the Quaternary tectonics [also see 47], suggesting that GPS velocity fields can be used to infer long-wavelength deformation (>500 km) patterns over geological time scales. The work of [46] and [45] also shows that northern Tibet is undergoing counterclockwise rotation whereas southern Tibet is undergoing clockwise rotation. In detail, the velocity component in the N20°E direction (relative to Siberia) decreases linearly and the velocity component in the N110°E direction is parabolic in N-S profiles. The maximum value of the eastward velocity component increases eastward as a result of east-west extension. To illustrate the above points more clearly and to search for the physical cause of the observed velocity field, we express the Tibetan GPS velocity distribution in the following analytical forms

$$u = Ax + Cy^2 + C_1, \ \ v = -By + C_2 \quad (7)$$

and the resulting strain-rate field as

$$\dot{\varepsilon}_{xx} = A, \quad \dot{\varepsilon}_{yy} = -B, \quad \dot{\varepsilon}_{xy} = Cy \tag{8}$$

where A, B, C, C_1 and C_2 are positive constants, the x and y axes point to N20ºE and N110ºE, the y axis lies along the dividing line separating the northern and southern domain of the Tibetan conjugate fault zone, and u and v are velocity components in the x and y directions. Constants C_1 and C_2 in the above equations represent rigid-body translation while the rotation field (i.e., vorticity) can be obtained by the following,

$$\omega_z = \frac{\partial v}{\partial x} - \frac{\partial u}{\partial y} = 2Cy \tag{9}$$

The above relationship indicates that the positive (northern) quadrants experiences counterclockwise rigid-body rotation while the negative (southern) quadrants experiences clockwise rigid-body rotation; the magnitude of rotation increases away from x axis at $y = 0$. The sense of rigid-body rotation is consistent with the presence of two shear zones having opposite sense of shear: left-slip in the north and right-slip in the south. Equation (8) captures this kinematic property as $\dot{\varepsilon}_{xy} = Cy$ changes its sign when y changes the sign across the dividing line between the northern left-slip and southern right-slip fault domains of the Tibetan conjugate fault system. Because the eastward velocity component varies *nonlinearly* northward, the two shear zones are not simple-shear zones in a strict sense as the eastward velocity component is not a linear function of a northward distance and the shear zone is experiencing perpendicular contraction. Nevertheless, [46] obtained the average right-slip rate in the south as 10 ± 2 mm/yr and the average left-slip rate in the north as 9.0 ± 1.5 mm/yr.

From the Tibetan GPS data, we hypothesize that collision-induced compression may have produced *an orogen-parallel flow consisting of two oppositely moving simple-shear zones*. Because simple-shear deformation generates Riedel shears at ~15-25° from the main shear direction [e.g., 41, 48], development of the two parallel shear zones can lead to the formation of paired Riedel shears with opposite shear sense at 65-75° from the maximum regional compressive direction. We note that the eastward GPS velocity distribution in central Tibet is similar to a Hagen-Poiseuille flow field expressed as the following for a Newtonian fluid [49]

$$u = \frac{1}{8} G(4y^2 - h^2) \tag{10}$$

where $G = \frac{dp}{dx}$ is the pressure gradient in the u direction, h the width of the channel, and y the direction perpendicular to the flow. The resulting shear strain-rate distribution is

$$\dot{\varepsilon}_{xy} = \frac{1}{2}\frac{\partial u}{\partial y} = \frac{1}{2} Gy, \tag{11}$$

which is comparable to the shear strain distribution of $\dot{\varepsilon}_{xy} = Cy$ across central Tibet. A key difference between the distribution of the Tibetan eastward velocity and a Hagan-Poiseuille flow is that the former increases in magnitude eastward due to east-west extension while the latter maintains a constant velocity in the flow direction.

The similarity between the Tibetan GPS velocity field and the Hagen-Poiseuille flow raises the question of what generates the pressure gradient. Dynamic modeling indicates that the active tectonics of Tibet could be driven by gravitational-gradients [50], topographic load [32], laterally varying boundary force [51], laterally changing viscosity [52, 53], shear heating [54] and focused erosion [e.g., 55]. We show via sandbox experiments that subhorizontal shear could be another driving mechanism for Tibetan tectonics. Our experimental apparatus allows the channel width to adjust with movable sidewalls. We simulate basal shearing by sliding a thin plate, producing parallel shear zones above. The experimentally created fault pattern with paired Riedel shears is similar to the observed V-shaped conjugate systems. We found no deformation above the sliding plate if the underlying channel is too narrow or basal friction is too low. Interestingly, Riedel shears developed first near the fore edge of the channel are rotated the most outward, with the sense of rotation opposite to those predicted by the vertical-axis rotation models mentioned above, but is consistent with the present-day strain-rate field. The displacement in our experiments is parabolic and increases in the flow direction, similar to the eastward GPS velocity distribution, implying that lower-crustal and upper mantle processes may play an active role in Tibetan deformation.

Our proposed research examines the role of strain state rather than stress state for fault formation and development. This new perspective leads to a general paired-shear-zone model that may explain the formation of all three-types of non-Andersonian conjugate faults, including the low-angle normal faults and high-angle reverse faults [e.g., 5, 38, 56]. Our model contrasts sharply to the Anderson fault theory that emphasizes the role of a simple and uniform state of stress as the sole factor in affecting fault formation. Our new model opens the possibility that the same fault pattern can be related to a unique velocity field generated by diverse dynamic mechanisms, ranging from laterally varying gravitational or boundary forces, focused erosion, to basal shearing, to mention a few. Our proposed research links upper crustal deformation in collisional orogens to subhorizontal flow in the lower crust and upper mantle. This has implications for the fundamental nature and dynamic cause of continental deformation.

2. References

[1] Anderson E M 1951 *The Dynamics of Faulting and Dyke Formation with Applications to Britain* (Edinburgh: Oliver and Boyd)
[2] Freund R 1970 *J. Geol.* **78** 188–200
[3] Sylvester A G 1988 *GSA Bull.* **100** 1666–1703
[4] Price R A 1981 The Cordilleran foreland thrust and fold belt in the southern Canadian Rocky Mountains *Thrust and Nappe Tectonics* eds M P Coward and K R McClay (Geol. Soc. Lond Special Publication) **9** 427–48
[5] Lister G S and Davis G A 1989 *J. Struct. Geol.* **11** 65–94
[6] Wernicke B 1992 Cenozoic extensional tectonics of the U.S. Cordillera *The Geology of North America G-3* eds B C Burchfiel, P W Lipman and M L Zoback (Boulder, Colorado: Geological Society of America) pp 553–81
[7] Ratschbacher L, Merle O, Davy P and Cobbold P 1991 *Tectonics* **10** 245–56
[8] Ratschbacher L, Frisch W, Linzer H G and Merle O 1991 *Tectonics* **10** 257–71
[9] Sengor A M C and Kidd W S F 1979 *Tectonophysics* **55** 361–76
[10] Jackson J and McKenzie D 1984 *Geophys. J. Royal Astron. Soc.* **77** 185–264
[11] Dhont D, Chorowicz J and Luxey P 2006 Anatolian escape tectonics driven by Eocene crustal thickening and Neogene-Quaternary extensional collapse in the eastern Mediterranean region *Postcollisional tectonics and magmatism in the Mediterranean region and* Asia eds Y Dilek and S Pavilides S (Geological Society Special Paper) **409** 331–462
[12] Tapponnier P, Mattauer M, Proust F and Cassaigneau C 1981 *Earth Planet. Sci. Lett.* **52** 335–71
[13] Brookfield M E and Hashmat A 2001 *Earth Sci. Rev.* **55** 41–71
[14] Yin A 2000 *J. Geophys. Res.* **105** 21745–59
[15] Taylor M, Yin A, Ryerson F, Kapp P and Ding L 2003 *Tectonics* **22** doi:10.1029/2002TC001361
[16] Taylor M and Peltzer G *2006 J. Geophys. Res* **111** doi:10.1029/2005JB004014
[17] Cunningham D 2005 *Earth Planet. Sci. Lett.* **240** 436–44
[18] Walker R T, Nissen E, Molor E and Bayasgalan A 2007 *Geology* **35** 759–62

[19] Leloup P H, Lacassin R, Tapponnier R, Zhong D, Lui X, Zhang L and Ji S 995 *Tectonophysics* **251** 3–84
[20] Morley C K 2001 *J. Geol. Soc.* **158** 461–74
[21] Morley C K, Woganan N, Sankumarn N, Hoon T B, Alief A and Simmon M 2001 *Tectonophysics* **334** 115–50
[22] Kornsawan A and Morley C K 2002 *J. Struct. Geol.* **24** 435–90
[23] Backe G, Dhont D and Hervouet Y 2006 *Tectonophysics* **425** 25–53
[24] Escalona A and Mann P 2006 *AAPG Bull.* **90** 657–78
[25] Molnar P and Tapponnier P 1975 *Science* 189 419–26
[26] Tapponnier P, Peltzer G, Le Dain A Y, Armijo R and Cobbold P 1982 *Geology* **10** 611–16
[27] Peltzer G and Tapponnier P 1988 *J. Geophys. Res.* **93** 15,085–117
[28] Peltzer G 1988 Bulletin Geological Institute, University of Uppsala **14** 115–28
[29] Davy P and Cobbold P 1988 Bulletin Geological Institute University of Uppsala **14** 129–41
[30] England P and Houseman G 1986 *J. Geophys. Res.* **91** 3664–76
[31] Avouac J P and Tapponnier P 1993 *Geophys. Res. Lett.* **20** 895–8
[32] Clark M K and Royden L H 2000 *Geology* **28** 703–6
[33] Bendick R and Flesch L 2007 *Geology* **35** 895–8
[34] Ron H, Freund R, Garfunkel Z and Nur A 1984 *J. Geophys. Res.* **89** 6256–70
[35] McKenzie D and Jackson J 1986 *J. Geol. Soc. London* **143** 349–53
[36] Dewey J F, Cande S and Pitman W C 1989 *Eclogae Geologicae Helvetiae* **82** 717–34
[37] Donath F A 1961 *GSA Bull.* **72** 985–90
[38] Sibson R H, Robert F and Poulsen K H 1988 *Geology* **16** 551–5
[39] Yin Z M and Ranalli G 1992 *J. Struct. Geol.* **14** 237–44
[40] Jeager J C and Cook N G W 1979 *Fundamentals of Rock Mechanics, 3rd ed* (London: Chapman and Hall, London)
[41] Tchalenko J S 1970 *GSA Bull* **81** 41–60
[42] Wilcox R E, Harding T P and Seely D R 1973 *AAPG Bull.* **57** 74–96
[43] Eisenstadt G and Sims D 2005 *J. Struct. Geol.* **27** 1399–412
[44] Yin A and Harrison T M 2000 *Ann. Rev. Earth Planet. Sci.* **28** 211–80
[45] Allmendinger RW, Reilinger R and Loveless J 2007 *Tectonics* **26** TC3013 doi:10.1029/2006TC002030
[46] Zhang P, Shen Z, Wang M, Gan W, Burgmann R and Molnar P 2004 *Geology* **32** 809–12
[47] England P and Molnar P 2005 *J. Geophys. Res.* **109** B12401 doi: 10.1029/2004JB00354
[48] Naylor M A, Mandl G and Sijpesteijn C H K 1986 *J. Struct. Geol.* **8** 737–52
[49] White F M 1986 *Fluid Mechanics, 2nd ed* (Boulder: McGraw-Hill)
[50] Flesch L M, Holt W E, Silver P G, Stephenson M, Wang C Y and Chan W W 2005 *Earth Planet. Sci. Lett.* **238** 248–68
[51] Kapp P and Guynn J H 2004 *Geology* **32** 993–6
[52] Liu M and Yang Y Q 2003 *J. Geophys. Res.* **108** Art. No. 2361
[53] Copley A and McKenzie D 2007 *Geophys. J. Int.* **169** 683–98
[54] Jimenez-Munt I and Platt J P 2006 *Tectonics* **25** TC6002 doi:10.1029/2006TC001963
[55] Beaumont C, Jamieson R A, Nguyen M H and Lee B 2001 *Nature* **414** 738–42
[56] Yin A 1989 *Tectonics* 8 469–82

IOP Conf. Series: Earth and Environmental Science **2** (2008) 012027 doi:10.1088/1755-1307/2/1/012027

Geomorphic response of an active metamorphic core-complex in a collisional orogen: Example from the Lunggar Shan, Southern Tibet

Michael H Taylor[1,3], Paul A Kapp[2] and Daniel F Stockli[1]
[1]Department of Geology, University of Kansas, 1475 Jayhawk Blvd., Lawrence, KS 66044, USA
[2]Department of Geosciences, University of Arizona, Tucson, AZ, USA

E-mail: mht@ku.edu

Abstract. We present structural and neotectonic mapping from the Lunggar Shan rift in southern Tibet. The Lunggar Shan is a N-trending mountain range ~70 km long N-S and up to 40 km wide E-W. The Lunggar Shan is bounded on its east side by a low-angle (<40°) east-dipping detachment fault that juxtaposes mylonitic gneiss and variably deformed granites in its footwall against alluvial fans and Neogene gravels in its hangingwall. Foliations in the mylonitic footwall dip <40° east and stretching lineations are east plunging. The range front detachment is presently inactive as indicated by undisturbed moraines and Quaternary sediments that overlie it. However, we consider the Lunggar Shan detachment to be an active structure, as inferred by range parallel fault scarps cutting Quaternary alluvium located 4-5 km into the hangingwall basin, with >40 m of throw on individual scarps. An intriguing observation is that an intrabasinal topographic high is actively developing near areas of inferred maximum extension, with lacustrine sediments being uplifted and eroded. This observation indicates that the rift basin initially developed as a typical half-graben system that underwent a transition from deposition, to uplift and erosion perhaps as a result of isostatic rebound of the footwall at depth, warping the overlying hangingwall basin. If correct, the Lunggar Shan may represent a modern analogue to the supradetachment basin model.

1. Introduction

It has long been speculated and predicted from geodynamic models that the development of low-angle normal fault (detachment) systems and extensional metamorphic core complexes is favored in high-elevation regions characterized by hot, thick crust [1, 2, 3, 4, 5]. Furthermore, the presence of a weak mid-crustal layer, decoupled from the upper crust and capable of flow, has been invoked to explain some geometric aspects of core complex activation of slip on detachments in the brittle upper crust [6, 7, 8, 9, 10, 11, 12, 13, 14, 15]. Despite these widely held views, geological evidence demonstrating a direct linkage between the "core complex" mode of extension and thickened crust is scarce because the best studied detachment fault systems that are proposed to be presently active are located at low elevations and characterized by thin crust (e.g., in Papua New Guinea, Baja California, Salton Trough, Death Valley, Mediterranean) [16, 17, 18, 19, 20, 21, 22, 23, 24]. The only well documented exception, to our knowledge, is the Cordillera Blanca detachment fault system within the Peruvian

[1] To whom any correspondence should be addressed.

Andes, which is anomalous in that its formation is attributed to the combined affects of thickened crust and the subduction of a buoyant oceanic ridge [25, 26].

The ideal natural laboratory to investigate the mode of extension within the ***hinterlands*** of collisional orogens characterized by anomalously hot and thick crust is in southern Tibet where ~N-trending rifts are actively developing above an extremely thick crust (65-85 km) [27, 28, 29] that arguably includes a fluid-bearing mid-crustal layer capable of flow [30, 31, 32, 33, 34, 35, 36, 37, 38]. Our understanding of the history of deformation and basin development associated with rifts in Tibet remains infant, yet recent studies reveal some common observations. First, Tibetan rifts are widely spaced (~100-300 km) [39], which may reflect mantle involvement, yet characterized by narrow-width rift flank uplifts (30-40 km) and basins (<10-15 km), consistent with isostatic compensation occurring in a low-viscosity lower crust [40]. Nascent rifts that have accommodated minimal (a few km's) extension are characterized by range-bounding high-angle normal faults and basins that show minimum basin elevations and depocenters in the central part of the rift [41]. In contrast, Tibetan rifts with more than ~15km of extension appear to be characterized by range-bounding, low-angle normal faults, structurally underlying extensional shear zones, and intrabasin topographic highs where basin fill is actively being uplifted, incised and which correspond to areas of inferred maximum extension [41]. Our hypothesized kinematic history calls upon an evolving breakaway fault system that leads to active slip on detachment faults in the uppermost crust and uplift and erosion of rift basin fill above the detachments in response to tectonic unloading and associated isostatic rebound of the footwall at depth. In contrast to North American Cordilleran detachment fault systems, extension is localized near rift-bounding faults (i.e., the hanging wall remains unextended outside of rift).

2. Active Deformation in Central Tibet

The Himalayan-Tibetan orogen is the archetype example of an active continent-continent collision zone and has served as the premier natural laboratory for investigating how the continental lithosphere deforms and high-elevation plateaus develop during collisional orogenesis. It has been known for decades that much of Tibet is currently undergoing ~E-W extension [42, 43, 44, 45, 46, 47, 48]. This extensional system is impressive in spatial extent, relief (locally >2 km), and rate (~2 cm/yr, one half the rate of Indo-Asian convergence). Numerous models have attempted to explain the tectonic significance of this syncollisional extension, generally calling upon one or some combination of the following: gravitational collapse due to crustal thickening or following removal of mantle lithosphere [49, 50], oblique convergence at the Himalayan front and growth of the Himalayan arc [46, 47, 51], eastward lateral extrusion of central Tibet with respect to southern Tibet [52-54], distributed eastward extrusion of crustal wedges [55, 56], and basal shear tractions related to underthrusting of Indian lithosphere [57]. Several studies have sought to constrain the initiation age of extension to test the aforementioned models or to date when Tibet supposedly achieved its maximum sustainable elevation [58, 59, 60, 61, 62, 63, 64, 65, 66].

3. Structural Geology of the Lunggar Rift

The N-trending Lunggar Range parallels a rift valley in the northern Lhasa terrane of west-central Tibet. The Lunggar Rift valley is ~70 km long and ~5-10 km wide [41]. The central part of the Lunggar Rift is bounded on its western flank by a <40° E-dipping normal fault that juxtaposes variably deformed biotite granite, leucogranite, and mylonitic gneiss in the footwall against Paleozoic strata and Neogene alluvial fan, fluvial and lacustrine rocks in the hanging wall. The proximal footwall includes chloritized breccia, cataclasites, and mylonitic gneisses with foliations subparallel to the normal fault and E-plunging stretching lineations and S-C fabrics indicating top-to-the-east sense-of-shear. We refer to this relatively low-angle normal fault with mylonitic rocks in the footwall as the "Lunggar detachment" [41]. Neogene gravels in the proximal hanging wall record the unroofing of the Lunggar Range as indicated by growth strata with clasts of both footwall and hanging wall lithologies. In turn, these gravels are cut by numerous N-striking, brittle normal faults. The brittle hanging-wall normal faults are both W- and E-dipping, and fault striae and sense-of-shear indicators indicate dominantly top-to-the-E displacement [41].

Presently, the Lunggar detachment is inactive at the surface as indicated by undisturbed moraines and alluvial fans that unconformably overlie it. However, ~4 km east of the inactive range front, ***active*** range-parallel E-dipping normal faults cut Quaternary alluvium within the rift basin. One example of active faulting in the central Lunggar basin is a flight of fluvial terrace risers developed along an E-flowing drainage that records the progressive incision of the river and associated abandonment of the higher terrace levels. At least five levels of terrace development record successive incision by the E–flowing stream. North-striking normal faults cut the terrace risers providing estimates of net throw that ranges from ~1m for the youngest surfaces to >35m for the oldest cumulative fault scarps. The central Lunggar Rift basin is bounded on its eastern margin by a system of high-angle W-dipping normal faults with Cretaceous strata in the footwall. Gently to moderately E-dipping rift basin fill occurs in the hanging wall and is well exposed due to recent incision. It consists of ~200 m of lacustrine mudstone capped by conglomerates derived from the Lunggar Range, based on abundant granitic and gneissic clasts [41].

The northern and southern Lunggar Rift is bounded on its western flank by high-angle E-dipping normal faults. Footwall rocks to the south include orthogneisses, although our preliminary studies suggest that the gneissic fabric predates normal faulting, whereas those to the north consist mainly of Paleozoic strata showing no any evidence of penetrative deformation. An intriguing observation is that normal fault scarps cutting Quaternary deposits in the northern and southern part of the Lunggar Rift are located along or within hundreds of meters of the range front, in contrast to the central part of the range where fault scarps are located kilometers from the range front. Preliminary apatite (U-Th)/He data from the central low-angle normal fault of the Lunggar Range footwall suggest the range underwent significant and rapid exhumation in response to low-angle extensional faulting in Plio-Pleistocene time. Ages cluster between 0.5-1.5 Ma [41].

4. Geomorphic Response of an Active Metamorphic Core-complex.

For the Lunggar Rift we investigated the spatial along-strike geomorphic response of the footwall and hanging wall by analyzing stream channel profiles. Because of the effects of lithologic and structural controls on stream channel profiles, we limited our analysis to the central region of the Lunggar Rift because of the uniform lithology and structural relationships. We subdivide individual along channel profiles into three regions based on the following geomorphic indices; an upper glacial dominated reach, an intermediate reach characterized by debris flow dominated incision with low channel concavity, and a downstream fluvial dominated reach with higher channel concavity. We observe channel concavities to increase (i.e., channel profile steepness decreases) southward along the central part of the Lunggar Range, with channel profiles to the north dominated by debris flow and glacial dominated reaches. These preliminary observations suggest that the Lunggar Range is likely a transient landscape responding to along-strike variations in fault geometry and magnitude of fault slip.

Another intriguing observation of the landscape is that a drainage divide occurs in the central part of the Lunggar Rift valley in the region of maximum inferred extension. To understand the spatial relationship between the hanging wall drainage divide and the footwall, rift parallel topographic swath profiles were generated by digitizing a polygon defined by the extent of Quaternary sediments in the basin, and a 4km swath spanning the crest of the Lunggar Range. The topographic data are then collapsed onto rift parallel profiles. When comparing elevation versus distance relative to the region of inferred maximum extension, we observe maximum basin elevations that coincide with the region of inferred maximum slip for the Lunggar detachment. This is in contrast to Tibet rifts with smaller magnitudes of displacement (< 5km's) which have minimum elevations occurring in the middle of the rift, consistent with a typical half graben geometry.

These observations led to the development of the following hypothesis for the evolution of the Tibetan grabens, where the rifts initiate with development of a breakaway fault as a typical half graben system with internally drained conditions located in the central region of the rift basin. As extension increases, the breakaway fault gets back rotated to a low angle, becomes abandoned and a new breakaway fault is initiated. With further tectonic unloading of the footwall at depth, the footwall undergoes isostatic rebound, resulting in upwarping of the detachment and uplift of the hanging wall basin and migration of the active depocenters toward the rift tips. With further extension, mylonitic rocks get exhumed to the surface and locally, the locus of active faulting steps basinward, resulting in basin cannibalism, and narrowing of the rift valley with progressive extension.

5. References

[1] Buck W R 1991 *Journal of Geophysical Research* **96** 20161–78
[2] Buck W R 1993 *Geology* **21** 933–6
[3] Coney P J and Harms T A 1984 *Geology* **12** 550–4
[4] Lachenbruch A H and Morgan P 1990 *Tectonophysics* **174** 39–62
[5] Regenauer-Lieb K, Weinburg R and Rosenbaum G 2006 *Nature* **442** 67–70
[6] Block L and Royden L H 1990 *Tectonics* **9** 557–67
[7] McKenzie D, Nimmo F, Jackson J A, Gans P B and Miller E L 2000 *Journal of Geophysical Research* **105** 11,029–46
[8] Spencer J 1984 *Geology* **12** 95–8
[9] Wdowinski S and Axen G J 1992 *Tectonics* **11** 303–15
[10] Wernicke B 1990 *The fluid crustal layer and its implications for continental dynamics* (Netherlands: Kluwer Academic Publishers) pp 509–544
[11] Wernicke B 1992 *Cenozoic extensional tectonics of the U.S. Cordillera* (Boulder, CO: Geological Society of America) pp 553–581
[12] Wernicke B and Axen G J 1988 *Geology* **16** 848–51
[13] Westaway R 1998 *Journal of the Geological Society* **155** 233–53
[14] Westaway R 1999 *Tectonophysics* **308** 407–43
[15] Yin A 1989 *Tectonics* **8** 469–82
[16] Abers G A, Ferris A, Craig M, Davies H, Lerner-Lam A L, Mutter J C and Taylor B 2002 *Nature* **418** 862–5
[17] Abers G A and Roecker S W 1991 *Journal of Geophysical Research* **96** 6379–401
[18] Axen G J, Fletcher J M, Cowgill E, Murphy M, Kapp P, MacMillan I and Ramos-Velazquez E 1999 *Geology* **27** 247–50
[19] Baldwin S L, Lister G S, Hill E J, Foster D A and McDougall I 1993 *Tectonics* **12** 611–28
[20] Cowan D S, Cladouhos T T, Morgan J K 2003 *Geological Society of America Bulletin* **115** 1230–48
[21] Hayman N W, Knott J R, Cowan D S, Nemser E and Sarna-Wojcicki A M 2003 *Geology* **31** 343–6
[22] Lister G S 1984 *Geology* **12** 221–5
[23] Numelin T, Kirby E, Walker J D, Didericksen B 2007 *Geosphere* **3** 163–76
[24] Sorel D 2000 *Geology* **28** 83–6
[25] Giovanni M 2007 *Tectonic and thermal evolution of the Cordillera Blanca detachment system, Peruvian Andes: Implications for normal faulting in a contractional orogen* University of California PhD Thesis
[26] McNulty B and Farber D 2002 *Geology* **30** 567–70
[27] Kind R, Yuan X, Saul J, Nelson D, Sobolev S V, Mechie J, Zhao W, Kosarev G, Ni J, Achauer U and Jiang M 2002 *Science* **298** 1219–21
[28] Owens T J and Zandt G 1997 *Nature* **387** 37–43
[29] W Zhao, J Mechie, L D Brown, J Guo, S Haines, T Hearn, S L Klemperer, Y S Ma, R Meissner, K D Nelson, J F Ni, P Pananont, R Rapine, A Ross and J Saul 2001 *Geophysical Journal International* **145** 486–98
[30] Brown L D, Zhao W, Nelson K D, Hauck M, Alsdorf D, Ross M, Cogan M, Clark M, Liu X and Che J 1996 *Science* **274** 1688–90

[31] Clark M K and Royden L H 2000 *Geology* **28** 703–6
[32] Francheteau J, Jaupart C, Jie S X, Wen-Hua K, De-Lu L, Jia-Chi B, Hung-Pin W and Hsia-Yeu D 1984 *Nature* **307** 32–6
[33] Klemperer S 2006 C*rustal flow in Tibet: geophysical evidence for the physical state of Tibetan lithosphere, and inferred patterns of active flow* (London: Geological Society, London, Special Publications) pp 39–70
[34] Nelson K D, Zhao W, Brown L D, Kuo J, Che J, Liu X, Klemperer S L, Makovsky Y, Meissner R, Mechie J, Kind R, Wenzel F, Ni J, Nabelek J, Leshou C, Tan H, Wei W, Jones A G, Booker J, Unsworth M, Kidd W S F, Hauck M, Alsdorf D, Ross A, Cogan M, Wu C, Sandvol E and Edwards M 1996 *Science* **274** 1684-8
[35] Pham V N, Boyer D, Therme P, Yuan X C, Li L and Jin G Y 1986 *Nature* **319** 310–4
[36] Royden L H, Burchfiel B C, King R W, Erchie W, Zhiliang C, Feng S and Yuping L 1997 *Science* **276** 788–90
[37] Shapiro N, Ritzwoller M and Molnar P 2004 *Science* **305** 233–6
[38] Unsworth M, Jones A, Wei W, Marquis S, Gokarn J, Spratt J, I-M team 2005 *Nature* **438** 78–81
[39] Yin A 2000 *Journal of Geophysical Research* **105** 21,745–59
[40] Masek J G, Isacks B L and Fielding E J 1994 *Tectonics* **13** 659–67
[41] Kapp P, Taylor M, Stockli D, Ding L 2008 *Geology* **36** 7–10; doi: 10 1130/G24054A 24051
[42] Armijo R, Tapponnier P, Mercier J L, Han T-L 1986 *Journal of Geophysical Research* **91** 13,803–72
[43] Dewey J F, Shackleton R M, Chang C F and Sun Y Y 1988 *Philosophical Transactions of the Royal Society of London Series a-Mathematical Physical and Engineering Sciences* **327** 379–413
[44] Klootwijk C T, Conaghan P J and Powell C M 1985 *Earth and Planetary Science Letters* **75** 167–83
[45] Mercier J L, Armijo R, Tapponnier P, Carey-Gailhardis E, Han T-L 1987 *Tectonics* **6** 275–304
[46] Seeber L and Armbruster J G 1984 *Tectonophysics* **105** 263–78
[47] Seeber L and Pecher A 1998 *Geology* **26** 791–4
[48] Tapponnier P and Molnar P 1977 *Journal of Geophysical Research* **82** 2905-30
[49] England P and Jackson J 1989 *Annual Review of Earth and Planetary Sciences* **17** 197–226
[50] Molnar P, England P and Martinod J 1993 *Reviews of Geophysics* **31** 357–96
[51] Klootwijk C T, Conaghan P J, Powell C Mca 1985 *Earth and Planetary Science Letters* **75** 167–83
[52] Armijo R, Tapponnier P, Mercier J L and Han T L 1986 *Journal of Geophysical Research* **91** 13,803–72
[53] Armijo R, Tapponnier P and Tonglin H 1989 *Journal of Geophysical Research* **94** 2787–838
[54] Peltzer G 1988 *Bulletin of the Geological Institutions of the University of Uppsala* **14** 115–28
[55] Taylor M and Peltzer G 2006 *Journal of Geophysical Research* **111** doi:10 1029/2005JB004014
[56] Taylor M, Yin A, Ryerson F, Kapp P and Ding L 2003 *Tectonics* **22** doi:10 1029/2002TC001361
[57] McCaffrey R and Nabelek J 1998 *Geology* **26** 691–4
[58] Blisniuk P M, Hacker B R, Glodny J, Ratschbacher L, Bi S W, Wu Z H, McWilliams M O, Calvert A 2001 *Nature* **412** 628–32
[59] Coleman M E 1996 *Geological Society of America Bulletin* **108** 1596–607
[60] Dewane T, Stockli D, Taylor M, Lee J and Ding L 2006 Timing of Cenozoic E-W Extension in the Tangra Yum Co-Kung Co Rift, south-central Tibet *EOS Trans. American Geophysical Union Fall Meeting Supplement* (San Francisco, CA, 11-15 December 2006) Abst. T34C-04
[61] Garzione C N, DeCelles P G, Hodkinson D G, Ojha T P and Upreti B N 2003 *Geological Society of America Bulletin* **115** 3–20
[62] Harrison T M, Copeland P, Kidd W S F and Lovera O M 1995 *Tectonics* **14** 658–76

[63] Maheo G, Leloup P H, Valli F, Lacassin R, Arnaud N, Paquette J L, Fernandez A, Haibing L, Farley K A and Tapponnier P 2007 *Earth and Planetary Science Letters* **256** 233–43
[64] Stockli D, Taylor M, Yin A, Harrison T M, D'Andrea J, Kapp P and Ding L 2002 Late Miocene-Pliocene inception of E-W extension in Tibet as evidenced by apatite (U-Th)/He data *Geological Society of America Abstracts with Programs* **34** 411
[65] Williams H, Turner S, Kelley S and Harris N 2001 *Geology* **29** 339–42
[66] Yin A, Kapp P, Murphy M, Manning C, Harrison T M, Grove M, Ding L, Deng X and Wu C 1999 *Geology* **27** 787–90

Relationships between crustal structure and extension in the Basin and Range Province and East Africa

G Randy Keller[1]
University of Oklahoma, School of Geology and Geophysics, Norman, Oklahoma, 73019 USA

E-mail: grkeller@ou.edu

Abstract. The Basin and Range Province of the western United States and northern Mexico is often cited as a classic example of a wide rift. It is also a region where metamorphic core complexes such as the ones observed in the Aegean region are observed. On the other hand, the eastern arm (Kenya rift) of the East African rift is considered to be the classic example of a continental rift, which is by some definitions narrow. In this paper, these two features are briefly compared in terms of crustal structure and associated manifestations of extension.

1. The Basin and Range Province Compared to the East African Rift

The Basin and Range Province of the western United States is an unusually wide continental rift zone, and its structure and evolution have been the topic of much debate. In the Basin and Range, large scale Cenozoic extension has exhumed midcrustal rocks (from 10-20 km depth) along low angle detachment faults to form classic metamorphic core complexes (MCC) [e.g., 1]. Similar features have subsequently been discovered around the world, and the Aegean region is a notable example. There is disagreement as to whether the Basin and Range should be considered to be a wide rift or simply a highly extended terrane. On the other hand, the East African rift zone is widely considered to be the classic example of a continental rift, which is a narrow zone of lithospheric-scale extension [e.g., 2].

In the portion of the Basin and Range that contains the most MMCs (figure 1), controlled-source seismic surveys and receiver function studies have delineated a relatively uniform crustal thickness (~30 km) despite varying magnitudes of crustal extension (<10-300 km) [3]. In addition, the velocity of the lower crust rarely exceeds 6.5 km/s indicating that mafic material is volumetrically insignificant. However, there are areas of thickened lower crust that correlate with regions of greatest extension (Death Valley, Lake Mead, and Colorado River extensional corridor) and the long wavelength (100-400 km) component of the Bouguer gravity anomaly field (figure 2). The seismic velocity and density of these lower crustal features indicate that they are either the result of the intrusion of intermediate composition rocks directly from the mantle or the result of lateral ductile flow [3]. Seismic velocities in the upper mantle range from 7.9 km/s in the central Basin and Range, to 7.8 km/s in the southern Basin and Range and metamorphic core complex belt, to 7.9 km/s under the Colorado Plateau. In contrast to the extreme amounts of extension and internal deformation of the Basin and Range, the Colorado Plateau has remained a stable, relatively rigid block. The crustal thickness is 40-50 km and the plateau is resistant to the extensional processes that have shaped the Basin and Range.

---

[1] To whom any correspondence should be addressed.

As point of comparison, mature results from the Kenya Rift International Seismic Project (KRISP) in the eastern arm of the East African rift (figure 3) provide a rather clear picture of lithospheric structure [e.g., 4, 5]. The KRISP results demonstrate that the Kenya rift is associated with sharply defined lithospheric thinning and very low (7.5 km/s under the Kenya dome) upper mantle velocities down to depths of over 150 km. Major axial variations in crustal and uppermost structure were also discovered and correlate well with variations in the amount of extension, the physiographic width of the rift valley, the regional topography, and the regional gravity anomalies (figure 4). Magmatic activity has certainly modified the crust but not to the extent believed prior to KRISP. This modification has taken the form of an underplated layer that constitutes the lowermost crust and local mafic intrusions beneath volcanic centers (figure 4) [6]. These observations strongly contrast with the Basin and Range Province where there is modest magmatic modification of the crust, and little correlation between crustal thickness and topography with amount of extension.

These observations drive home the point that extension affects the lithosphere in highly variable ways depending on factors such as the pre-existing state of the lithosphere and fertility of the upper mantle.

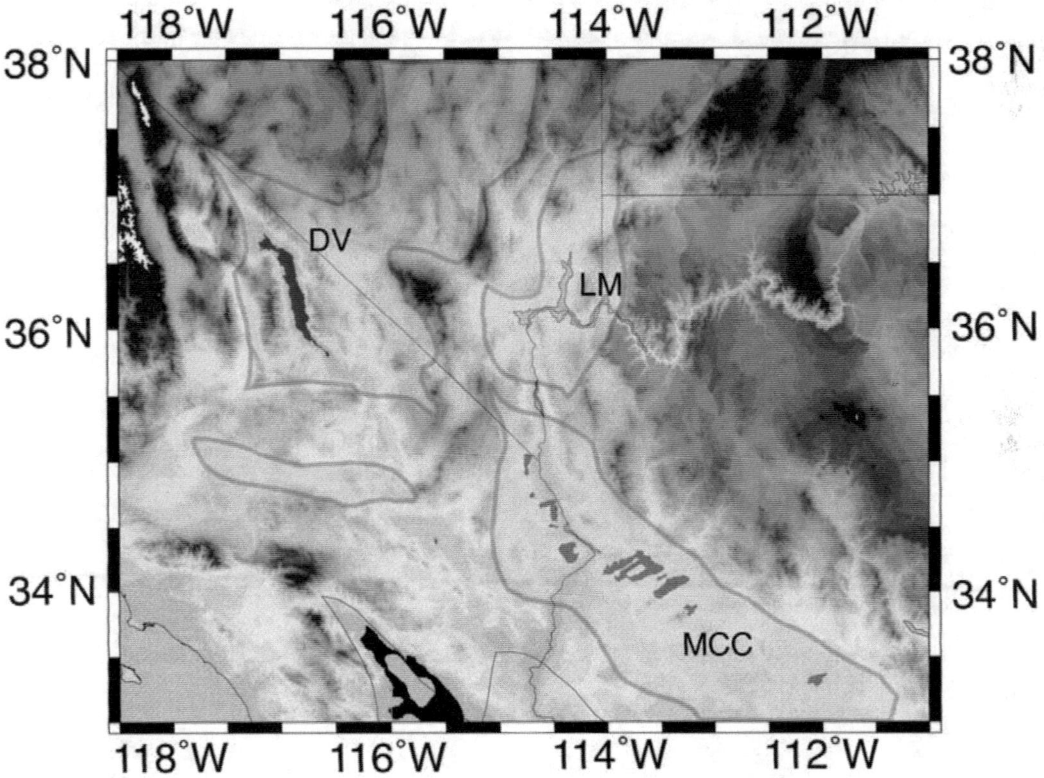

Figure 1. Index map of the region of the Basin and Range that has the greatest concentration of highly extended domains. Highly extended domains (DV - Death Valley; LM – Lake Mead; MCC – Metamorphic core complex belt; [1] are outlined in orange and the MCCs are filled with orange. Modified from [3].

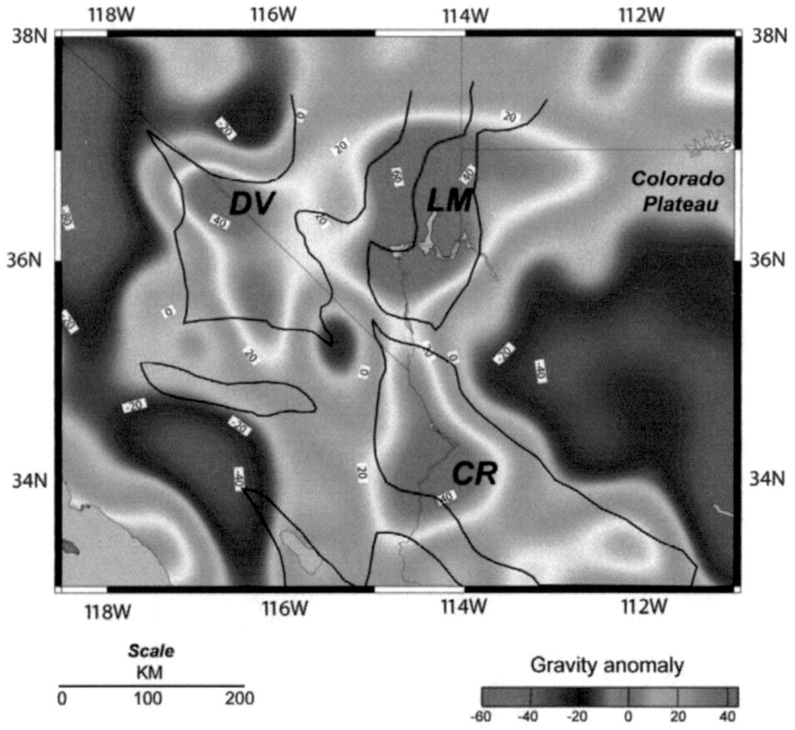

Figure 2. Gravity anomaly map showing the long wavelength (100-1500km) component of the Bouguer anomaly field. Modified from [3]. DV - Death Valley; LM – Lake Mead; CR- Colorado River extensional corridor from [1].

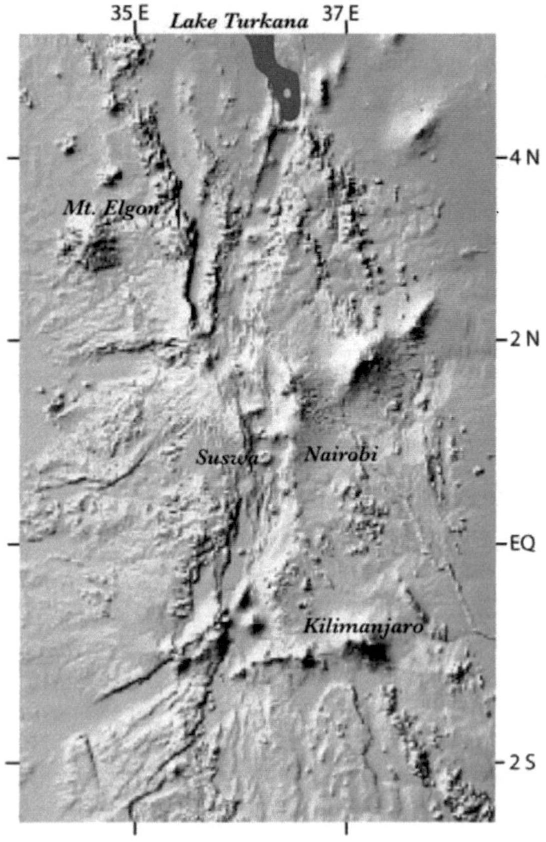

Figure 3. Digital elevation map of the Kenya rift area constructed from the Shuttle Radar Topography Mission (SRTM) data. Provided by NASA. The computer model shown in Figure 4 is of a N-S profile that extends along 36°E.

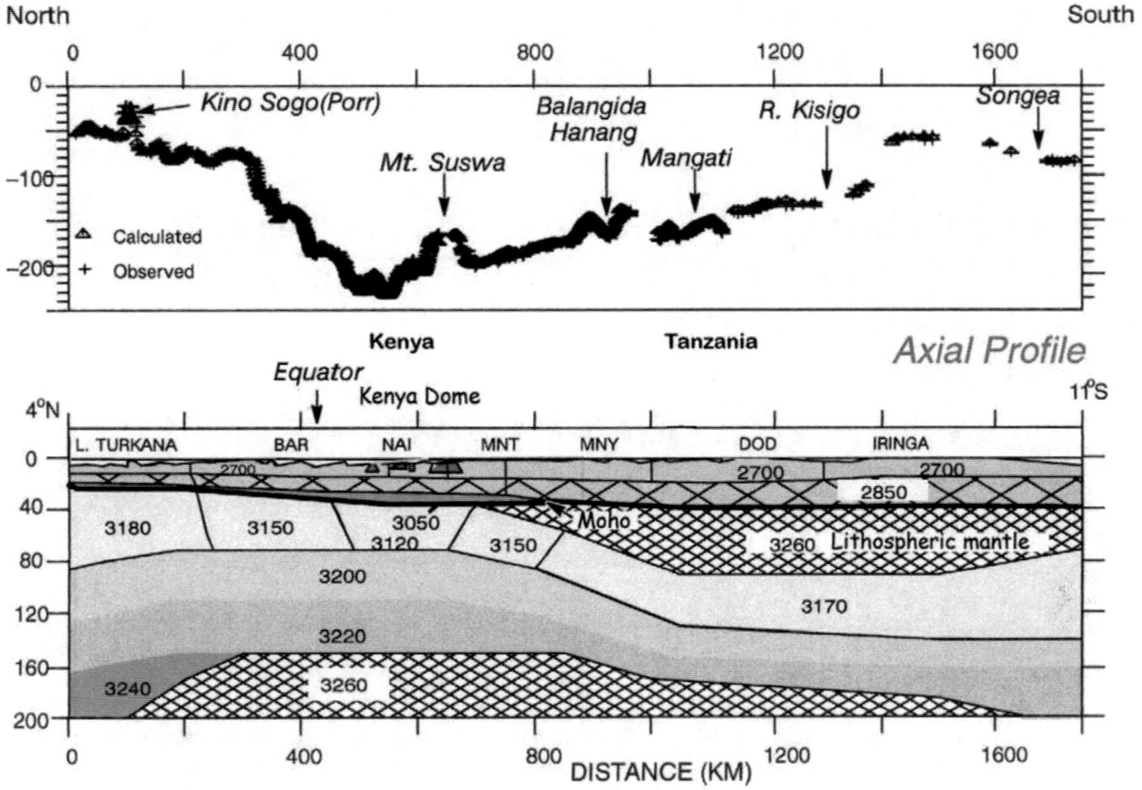

Figure 4. Integrated geophysical model along the axis of the Kenya rift. This model approximately follows 36ºE in Figure 3. BAR-Baringo-Bogoria basin; NAK-Nakuru-Naivasha basin; MNT-Magadi-Natron basin; MNY-Lake Manjara basin; DOD-Dodoma. Numbers indicate densities in kg/m^3. Modified from [6].

2. References

[1] Wernicke B P 1992 Cenozoic extensional tectonic of the U.S. Cordillera *The geology of North America, The Cordilleran Orogen, Conterminous U.S.* (vol G-3) eds B C Burchfiel, P W Lipman and M L Zoback (Boulder: Geological Society of America) pp 553–81

[2] Burke K 1980 Intracontinental rifts and aulacogens *Continental Tectonics* (Washington, D.C.: National Academy of Sciences) 42–9

[3] Hicks N O 2001 Lithospheric structure of the Basin and Range province – southwestern Colorado plateau: Southeastern California, southern Nevada, and western Arizona (University of Texas at El Paso: M.S. Thesis)

[4] Keller G R, Prodehl C, Mechie J, Fuchs K, Khan M A, Maguire P K H, Mooney W D, Achauer U, Davis P M, Meyer R P, Braile L W, Nyambok I O and Thompson G A 1994 The East African rift system in the light of KRISP 90 *Crustal and Upper Mantle Structure of the Kenya Rift* eds C Prodehl, G R Keller and M A Khan *Tectonophysics* **236** 465–83

[5] Prodehl C, Ritter J R, Mechie J, Keller G R, Khan M A, Jacob B, Fuchs K, Nyambok I O, Obel J D and Riaroh D 1997 The KRISP 94 lithospheric investigation of southern Kenya - the experiments and their main results *Tectonophysics* **278** 121–47

[6] Simiyu S M and Keller G R 1997 An integrated analysis of lithospheric structure across the East African plateau based on gravity anomalies and recent seismic studies *Tectonophysics* **278** 291–313

Death Valley turtlebacks: Mesozoic contractional structures overprinted by Cenozoic extension and metamorphism beneath syn-extensional plutons

Terry L Pavlis[1,3], **Marli Miller**[2] **and Laura Serpa**[1]
[1]Dept. of Geological Sciences, Univ. of Texas at El Paso, El Paso, TX 7996, USA
[2]Dept. of Geological Sciences, Univ. of Oregon, Eugene, OR 97403, USA

E-mail: tlpavlis@utep.edu

Abstract. The term turtleback was first coined to describe the curvilinear fault surfaces that produced a distinctive geomorphic form in the Black Mountains east of Death Valley, and although it was decades before their full significance was appreciated, they remain one of the most distinctive features of the extensional structure of the Death Valley region. Historically the interpretation of the features has varied markedly, and misconceptions about their character continue to abound, including descriptions in popular field guides for the area. It the 1990's, however, the full history of the systems began to be apparent from several key data: 1) the dating of the plutonic assemblage associated with the turtlebacks demonstrated that late Miocene, syn-extensional plutonism was fundamental to their formation; 2) the plutonic assemblage forms an intrusive sheet structurally above the turtlebacks, indicating a tie between much of the high grade metamorphism and Cenozoic plutonism; 3) a modern analog for the syn-extensional plutonism in the Black Mountains was recognized beneath Death Valley with the imaging of a mid-crustal magma body; 4) the Neogene structural history was worked out in the turtlebacks showing that folding of early-formed shear zones formed the turtleback anticlinoria but overprinting by brittle faults produced the final form as they cut obliquely across the older structure; and 5) the pre-extensional structural history was clarified, demonstrating that Mesozoic basement-involved thrust systems are present within the turtlebacks, but have been overprinted by the extensional system. An unresolved issue is the significance of Eocene U-Pb dates for pegmatites within the region, but presumably these relate somehow to the pre-extensional history. Miller and Pavlis (2005; E. Sci. Rev.) reviewed many features of the turtlebacks,and our working model for the region is that the turtlebacks originated as mid-crustal ductile-thrust systems within the Cordilleran fold-thrust belts. Our work to the east of Death Valley suggests these thrusts were part of a NW trending thrust system that overprinted an older NE trending fold-thrust system that tracks into the Death Valley region from Nevada. These NW trending thrusts probably underlie all of the southern Black Mountains (south of the turtlebacks) and we suggest that pre-extensional structural relief along these basement thrusts placed basement at shallow crustal levels throughout what is now the Black Mountains; a conclusion consistent with the absence of rocks younger than Cambrian beneath Tertiary unconformities throughout the southern Death Valley region. In Late Miocene time, a major detachment system formed and the turtlebacks represent a mid-crustal shear zone developed during that time period, but this system is older, and structurally beneath younger detachments systems that comprise the Amargosa fault system. During motion on the detachment, an ~2km thick plutonic sheet was emplaced along the shear zone forming the

[3] To whom any correspondence should be addressed.

Miocene plutonic assemblages of the Black Mountains, and produced upper amphibolite facies metamorphic assemblages along the floor of the pluton in what are now the Copper Canyon and Mormon Point turtlebacks, but the Badwater Turtleback escaped this metamorphism due to a different structural position. Motion continued along the floor of the pluton but syn-extensional folding produced structural relief along folds with axes parallel to the extension direction. Ultimately a new detachment system cut obliquely across the older extensional system, removing the roof of the pluton, but cutting down to its floor in the turtlebacks. This fault system formed a complex detachment system updip in the famous "Amargosa Chaos", and removing the entire cover sequence from the Black Mountains (~10-12km of crustal section). The turtlebacks are therefore a composite structure in which extension contemporaneous with folding, presumably as a result of distributed transcurrent motion during extension, was critical to their formation. In addition, syn-tectonic plutonism played a key role in strain localization and metamorphism within the assemblages that comprise the turtlebacks.

AUTHOR INDEX